Power Plant
# Performance Curves
### for Testing and Dispatch

by D. James Benton

# Preface

While certainly not unique to the power industry, performance curves are central to design, operation, and testing of modern power generation systems, as well as many similar industrial plants. We will develop and apply these important formulations to several types of such systems. We will also discuss and illustrate when these work well and when they don't. The most common use of these approximations is to correct test results and to predict the operational response of complex systems the environment in which they operate. Examples in the text are based on GateCycle™; although they apply equally well to any other heat balance code.[1] A mixture of English (U.S. Customary) and SI units are used throughout this text, as is common in the industry.

*All of the examples contained in this book,*
*(as well as a lot of free programs) are available at...*
*https://www.dudleybenton.altervista.org/software/index.html*

---

[1] GateCycle™ is a thermodynamic cycle modeling (i.e., heat balance) tool sold by General Electric. There are several other commercially available tools, including EBSILON® (sold by STEAG) and GT-Pro® (sold by ThermoFlow, Inc.)

i

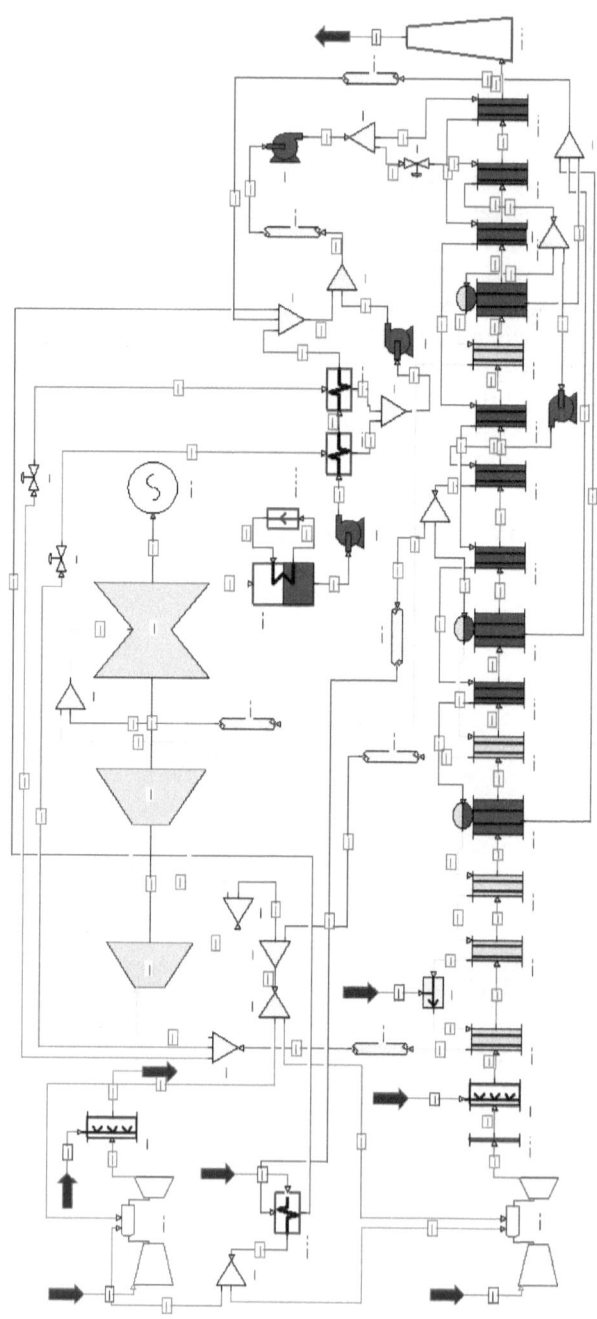

**Figure 1. Typical 2x1 Combined Cycle Power Plant in GateCycle™**

## Typical 2x1 Combined Cycle System Plant Corrections - without Duct Firing

| Additive Net Power Corrections | Units | A0 | A1 | A2 | A3 | A4 |
|---|---|---|---|---|---|---|
| Δ₁ Correction for HP Evaporator Blowdown | kW | -1.31024E+02 | 1.31024E-04 | 0 | 0 | 0 |
| Δ₂ Correction for Cooling Tower Wet-Bulb Difference at 99°F | kW | 0.00000E+00 | 2.98656E-01 | -7.51414E-01 | 0 | 0 |
| Δ₃ Correction for Cooling Tower Wet-Bulb Difference at 79°F | kW | 0.00000E+00 | -5.51120E-01 | 1.42886E-01 | 0 | 0 |
| Δ₄ Correction for Cooling Tower Wet-Bulb Difference at 59°F | kW | 0.00000E+00 | 2.12236E-01 | -3.82483E-02 | 0 | 0 |
| Δ₅ Correction for Cooling Tower Wet-Bulb Difference at 39°F | kW | 0.00000E+00 | -1.18797E+00 | 9.33919E-02 | 0 | 0 |
| Δ₆ Correction for Cooling Tower Wet-Bulb Difference at 19°F | kW | 0.00000E+00 | -2.61064E-01 | 6.31907E-01 | 0 | 0 |
| **Multiplicative Net Power Corrections** | **Units** | **A0** | **A1** | **A2** | **A3** | **A4** |
| α₁ Correction for Ambient Temperature | - | 8.72070E-01 | 2.75715E-03 | -3.03412E-05 | 4.59060E-07 | -1.93158E-09 |
| α₂ Correction for Barometric Pressure | - | 2.94953E+00 | -1.99676E-01 | 4.56158E-03 | 0 | 0 |
| α₃ Correction for Relative Humidity at 99°F | - | 9.41583E-01 | 8.70431E-02 | 1.71987E-02 | 0 | 0 |
| α₃ Correction for Relative Humidity at 79°F | - | 9.78116E-01 | 3.52129E-02 | 2.10081E-03 | 0 | 0 |
| α₃ Correction for Relative Humidity at 59°F | - | 9.94303E-01 | 9.15261E-03 | 5.70415E-04 | 0 | 0 |
| α₃ Correction for Relative Humidity at 39°F | - | 9.99641E-01 | 5.78389E-04 | 3.40170E-05 | 0 | 0 |
| α₃ Correction for Relative Humidity at 19°F | - | 1.00094E+00 | -1.54441E-03 | -4.25605E-05 | 0 | 0 |
| α₄ Correction for Fuel Temperature | - | 1.00039E+00 | -6.46914E-06 | -1.25334E-17 | 0 | 0 |
| α₅ Correction for LHV at C/H=4.068 | - | 9.85549E-01 | 6.12203E-07 | 0 | 0 | 0 |
| α₅ Correction for LHV at C/H=3.935 | - | 9.87507E-01 | 5.97335E-07 | 0 | 0 | 0 |
| α₅ Correction for LHV at C/H=3.781 | - | 9.89624E-01 | 5.82970E-07 | 0 | 0 | 0 |
| α₇ Correction for Fired Hours | - | 2.73851E-04 | 3.27203E-01 | 0 | 0 | 0 |
| **Multiplicative Heat Input Corrections** | **Units** | **A0** | **A1** | **A2** | **A3** | **A4** |
| β₁ Correction for Ambient Temperature | - | 8.45003E-01 | 3.95372E-03 | -5.78559E-05 | 7.87389E-07 | -3.18465E-09 |
| β₂ Correction for Barometric Pressure | - | 2.98612E+00 | -2.01753E-01 | 4.53352E-03 | 0 | 0 |
| β₃ Correction for Relative Humidity at 99°F | - | 9.30334E-01 | 1.01217E-01 | 2.48220E-02 | 0 | 0 |
| β₃ Correction for Relative Humidity at 79°F | - | 9.73867E-01 | 4.14315E-02 | 3.53818E-03 | 0 | 0 |
| β₃ Correction for Relative Humidity at 59°F | - | 9.92792E-01 | 1.18330E-02 | 3.00723E-04 | 0 | 0 |
| β₃ Correction for Relative Humidity at 39°F | - | 9.99430E-01 | 8.69585E-04 | 1.35272E-04 | 0 | 0 |
| β₃ Correction for Relative Humidity at 19°F | - | 1.00101E+00 | -1.70329E-03 | 4.33884E-05 | 0 | 0 |
| β₄ Correction for Fuel Temperature | - | 9.99961E-01 | 6.46914E-37 | -9.99092E-18 | 0 | 0 |
| β₅ Correction for LHV at C/H=4.068 | - | 9.72516E-01 | 1.25332E-06 | 0 | 0 | 0 |
| β₅ Correction for LHV at C/H=3.935 | - | 9.73663E-01 | 1.25923E-06 | 0 | 0 | 0 |
| β₅ Correction for LHV at C/H=3.781 | - | 9.75059E-01 | 1.26352E-06 | 0 | 0 | 0 |
| β₇ Correction for Fired Hours | - | 2.46480E-04 | 3.34077E-01 | 0 | 0 | 0 |

Figure 2. Table of Coefficients for Unfired Corrections

| Typical 2x1 Combined Cycle System Plant Corrections - with Duct Firing | | | | | | |
|---|---|---|---|---|---|---|
| **Additive Net Power Corrections** | Units | A0 | A1 | A2 | A3 | A4 |
| $\Delta_3$ Correction for HP Evaporator Blowdown | kW | -1.74877E+02 | 1.74877E+04 | 0 | 0 | 0 |
| $\Delta_{5a}$ Correction for Cooling Tower Wet-Bulb Difference at 99°F | kW | 0.00000E+00 | 1.28184E+02 | 2.39893E+00 | 0 | 0 |
| $\Delta_{5a}$ Correction for Cooling Tower Wet-Bulb Difference at 79°F | kW | 0.00000E+00 | 5.94542E+01 | 1.35830E+00 | 0 | 0 |
| $\Delta_{5a}$ Correction for Cooling Tower Wet-Bulb Difference at 59°F | kW | 0.00000E+00 | 3.15688E+01 | 5.87611E-01 | 0 | 0 |
| $\Delta_{5a}$ Correction for Cooling Tower Wet-Bulb Difference at 39°F | kW | 1.84261E-16 | 7.38747E+00 | -4.37626E-01 | 0 | 0 |
| $\Delta_{5a}$ Correction for Cooling Tower Wet-Bulb Difference at 19°F | kW | 0.00000E+00 | 4.97020E+00 | 2.68252E-01 | 0 | 0 |
| $\Delta_{p/d_3}$ Correction for Duct Firing | kW | 0.00000E+00 | -1.10631E-02 | 0 | 0 | 0 |
| **Multiplicative Net Power Corrections** | Units | A0 | A1 | A2 | A3 | A4 |
| $\alpha_1$ Correction for Ambient Temperature | - | 8.76721E-01 | 2.64155E-03 | -2.80792E-05 | 4.08246E-07 | -1.54108E-09 |
| $\alpha_2$ Correction for Barometric Pressure | - | 2.75259E+00 | -1.77075E-01 | 3.93548E-03 | 0 | 0 |
| $\alpha_3$ Correction for Relative Humidity at 99°F | - | 9.36454E-01 | 8.99712E-02 | 2.65639E-02 | 0 | 0 |
| $\alpha_3$ Correction for Relative Humidity at 79°F | - | 9.76259E-01 | 3.66658E-02 | 4.83651E-03 | 0 | 0 |
| $\alpha_3$ Correction for Relative Humidity at 59°F | - | 9.93284E-01 | 1.06814E-02 | 8.51854E-04 | 0 | 0 |
| $\alpha_3$ Correction for Relative Humidity at 39°F | - | 9.99460E-01 | 8.71298E-04 | 4.87366E-05 | 0 | 0 |
| $\alpha_3$ Correction for Relative Humidity at 19°F | - | 1.00092E+00 | -1.69887E-03 | 2.85048E-04 | 0 | 0 |
| $\alpha_4$ Correction for Fuel Temperature | - | 1.00047E+00 | -7.75590E-06 | -1.35719E-17 | 0 | 0 |
| $\alpha_5$ Correction for LHV at C/H=4.068 | - | 9.87075E-01 | 5.45082E-01 | 0 | 0 | 0 |
| $\alpha_5$ Correction for LHV at C/H=3.935 | - | 9.88361E-01 | 5.56512E-07 | 0 | 0 | 0 |
| $\alpha_5$ Correction for LHV at C/H=3.781 | - | 9.90518E-01 | 5.34474E-07 | 0 | 0 | 0 |
| $\alpha_6$ Correction for Fired Hours | - | 2.51821E-04 | 3.27489E-01 | 0 | 0 | 0 |
| **Multiplicative Heat Input Corrections** | Units | A0 | A1 | A2 | A3 | A4 |
| $\beta_1$ Correction for Ambient Temperature | - | 8.57261E-01 | 3.66572E-03 | -5.36398E-05 | 7.25165E-07 | -2.95048E-09 |
| $\beta_2$ Correction for Barometric Pressure | - | 2.72769E+00 | -1.72779E-01 | 3.75839E-03 | 0 | 0 |
| $\beta_3$ Correction for Relative Humidity at 99°F | - | 9.36888E-01 | 9.27176E-02 | 2.07814E-02 | 0 | 0 |
| $\beta_3$ Correction for Relative Humidity at 79°F | - | 9.76278E-01 | 3.77756E-02 | 2.93437E-03 | 0 | 0 |
| $\beta_3$ Correction for Relative Humidity at 59°F | - | 9.93434E-01 | 1.08034E-02 | 2.33137E-04 | 0 | 0 |
| $\beta_3$ Correction for Relative Humidity at 39°F | - | 9.99459E-01 | 8.99058E-04 | 4.94010E-06 | 0 | 0 |
| $\beta_3$ Correction for Relative Humidity at 19°F | - | 1.00092E+00 | -1.54574E-03 | 2.30066E-05 | 0 | 0 |
| $\beta_4$ Correction for Fuel Temperature | - | 1.00000E+00 | 0.00000E+00 | 0.00000E+00 | 0 | 0 |
| $\beta_5$ Correction for LHV at C/H=4.068 | - | 9.75173E-01 | 1.13048E-06 | 0 | 0 | 0 |
| $\beta_5$ Correction for LHV at C/H=3.935 | - | 9.76087E-01 | 1.14337E-06 | 0 | 0 | 0 |
| $\beta_6$ Correction for LHV at C/H=3.781 | - | 9.77404E-01 | 1.14509E-06 | 0 | 0 | 0 |
| $\beta_7$ Correction for Fired Hours | - | 2.31596E-04 | 3.30518E-01 | 0 | 0 | 0 |

**Figure 3. Table of Coefficients for Fired Corrections**

# Table of Contents

page

Preface ..................................................................................................... i
Chapter 1. Correction Curve Methodology........................................... 1
Chapter 2. Gas Turbine Performance Curves........................................ 23
Chapter 3. Expected Performance........................................................ 37
Chapter 4. Observed GT Performance ................................................. 55
Chapter 5. Observed CCPP Performance.............................................. 65
Appendix A. Thermodynamic Properties of Moist Air........................ 77
Appendix B. Higher vs. Lower Heating Value ..................................... 81
Appendix C. C/H vs. H/C Ratio............................................................ 83
Appendix D. Generator Curves............................................................. 85
Appendix E. Multivariate Regression .................................................. 91
Appendix F. Risk of Icing.................................................................... 93
Appendix G. Gas Turbine Heat and Mass Balance.............................. 97
Appendix H. Apparent Temperature Dependence of Heating Values ............ 101

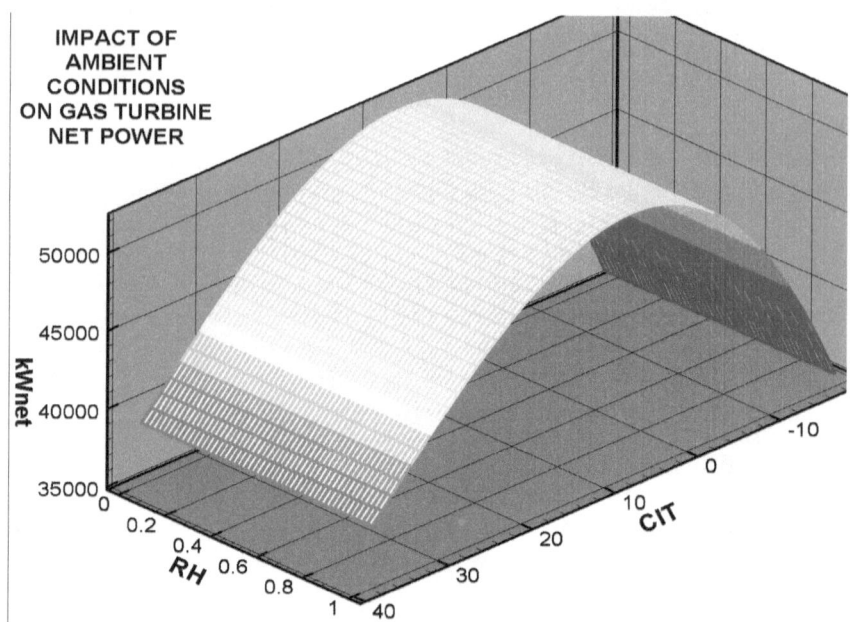

**Figure 4. Impact of Ambient Conditions on Gas Turbine Power**

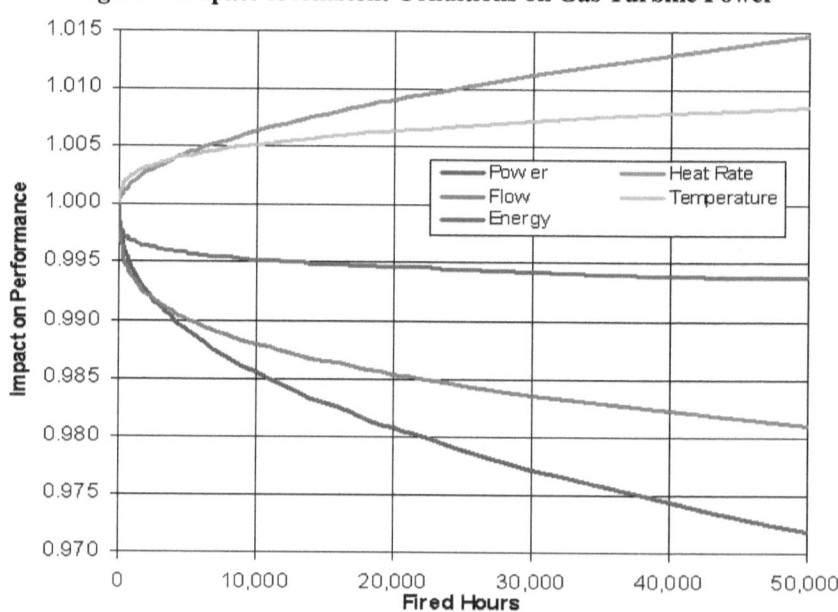

**Figure 5. Impact of Aging on Gas Turbine Performance**

# Chapter 1. Correction Curve Methodology

While the correction curve concept and methodology are by no means unique to the power industry or contractual acceptance testing, it essential to that process and we will begin there. The first test code we will consider is ASME PTC-46.[2] We will focus our discussion on Section 5 of this document, "Calculations and Results." Instruments and measurements are beyond the scope of this book, but not the Code. Here, we consider only calculations. Three equations are key. These provide guidance for calculating *corrected* power, heat input, and heat rate from *measured* values. The *measured* values are simply that—as obtained by calibrated precision instruments. The *corrected* values are that which would be expected were it possible to operate the system at all of the guarantee conditions simultaneously. The guarantee conditions include: atmospheric conditions (temperature, barometric pressure, humidity) as well as some operational conditions (grid frequency, power factor, and parasitic loads). The motivation for this calculation is simple: It is highly unlikely that a test might be conducted when all of these conditions were met.

$$P_{CORR} = \left( P_{MEAS} + \sum_{i=1}^{7} \Delta_i \right) \prod_{j=1}^{6} \alpha_j \qquad (1.1)$$

$$Q_{CORR} = \left( Q_{MEAS} + \sum_{i=1}^{7} \omega_i \right) \prod_{j=1}^{6} \beta_j \qquad (1.2)$$

$$HR_{CORR} = \frac{P_{CORR}}{Q_{CORR}} = \frac{\left( P_{MEAS} + \sum_{i=1}^{7} \Delta_i \right) \prod_{j=1}^{6} \alpha_j}{\left( Q_{MEAS} + \sum_{i=1}^{7} \omega_i \right) \prod_{j=1}^{6} \beta_j} \qquad (1.3)$$

We see that there are seven additive corrections ($\Delta_i$ and $\omega_i$) and six multiplicative corrections ($\alpha_j$ and $\beta_j$). While the ratio of the power ($\alpha$) and heat input ($\beta$) corrections is sometimes used ($f_j=\alpha_j/\beta_j$), we will devote an entire chapter to illustrating why this is not recommended. While these are presented in the context of overall plant performance (i.e., PTC-46), they apply equally well to individual gas turbines and a wide variety of devices. The corrections are explained in the following two tables.

---

[2] "Performance Test Code on Overall Plant Performance," American Society of Mechanical Engineers Performance Test Code No. 46, 1996.

## Table 1. Additive Corrections

| | |
|---|---|
| $\Delta_1/\omega_1$ | These terms account for energy flowing out of the system due to some external process. |
| $\Delta_2/\omega_2$ | These terms account for generator power factor or cooling. |
| $\Delta_3/\omega_3$ | These terms account for blowdown different from design.[3] |
| $\Delta_4/\omega_4$ | These terms account for energy flowing into the system due some external process, including makeup.[4] |
| $\Delta_{5A}/\omega_{5A}$ | These terms account for differing ambient conditions, such as the inlet of a cooling tower being different from that of a gas turbine in a combined cycle plant. |
| $\Delta_{5B}/\omega_{5B}$ | These terms account for cooling water supply (such as a pond, lake, or river) different from design. |
| $\Delta_{1C}/\omega_{5C}$ | These terms account for steam condenser pressure different from design and are only applicable if the condenser is removed from the test boundary. |
| $\Delta_6/\omega_6$ | These terms account for auxiliary loads different from design, for example, cooling tower fans or circulating water pumps. |
| $\Delta_7/\omega_7$ | This last category is most often used to account for duct firing. |

These additive corrections account for operational deviations from what is considered the normal range of conditions over which the system is expected to perform. These are non-routine adjustments—reasonable (non-extraordinary) but not standard. This quality distinguishes these from the next set.

## Table 2. Multiplicative Corrections

| | |
|---|---|
| $\alpha_1/\beta_1$ | These terms account for ambient temperature. |
| $\alpha_2/\beta_2$ | These terms account for barometric pressure. |
| $\alpha_3/\beta_3$ | These terms account for ambient humidity. |
| $\alpha_4/\beta_4$ | These terms account for fuel supply temperature. |
| $\alpha_5/\beta_5$ | These terms account for fuel analysis/composition. |
| $\alpha_6/\beta_6$ | These terms account for aging (i.e. equivalent hours of operation). |

---

[3] Blowdown is water or steam intentionally bled off of a process to be filtered or polished in order to avoid the accumulation of contaminants.

[4] Makeup is typically water (although it could be steam) fed back into a system (often after filtering) to conserve mass when some other flow is being removed.

These multiplicative corrections account for routine operation, that is, the normal range of conditions over which the system is expected to perform. These are not *deviations* and there should be no *surprises*. Snow in Minnesota (or Saskatchewan) is *not* a surprise! Any "plant" built above the 40th parallel is expected to operate even when it's snowing. I recall a plant built on the shore of the Dead Sea (the lowest spot on the face of the Earth and one of the hottest) that "stumbled" when the temperature reached 40°C. This is no surprise and is inexcusable from a design perspective.

Equations 1.1, 1.2, and 1.3 are cast so as to provide corrected (i.e., reference or base) results from measured. These can be inverted so as to provide expected from reference. Equations of the first type are most often called *correction* curves and of the second type are most often called *performance* curves.

$$P_{EXPECT} = \frac{P_{BASE}}{\prod\limits_{j=1}^{6} \alpha_j} - \sum_{i=1}^{7} \Delta_i \qquad (1.4)$$

$$Q_{EXPECT} = \frac{Q_{CORR}}{\prod\limits_{j=1}^{6} \beta_j} - \sum_{i=1}^{7} \omega_i \qquad (1.5)$$

$$HR_{EXPECT} = \frac{P_{EXPECT}}{Q_{EXPECT}} = \frac{\dfrac{P_{BASE}}{\prod\limits_{j=1}^{6} \alpha_j} - \sum\limits_{i=1}^{7} \Delta_i}{\dfrac{Q_{CORR}}{\prod\limits_{j=1}^{6} \beta_j} - \sum\limits_{i=1}^{7} \omega_i} \qquad (1.6)$$

The following are typical correction curves for a combined cycle power plant, that is, a generating station having one or more combustion gas turbines exhausting into a heat recovery steam generator (HRSG) to produce steam, which drives a turbine and generator to produce electrical power. Two delta curves (3 and 5A) are illustrated along with the first three multiplicative curves (i.e., ambient temperature, barometric pressure, and relative humidity).

Note that these corrections are most often given in U.S. Customary (i.e., English) units. The practicing scientist/engineer must be comfortable working with any and all systems of units. Arguments as to which units are preferable or superior rate as some of the most ridiculous and wasteful exchanges ever to occupy the inhabitants of Earth. My advice is: Get over it! We will use SI units in Chapter 2, where we focus specifically on gas turbines, and also Chapter 3.

3

Symbols indicate individual model run results. Curves indicate smooth regressions. Scatter is unavoidable due to the limited accuracy and convergence.

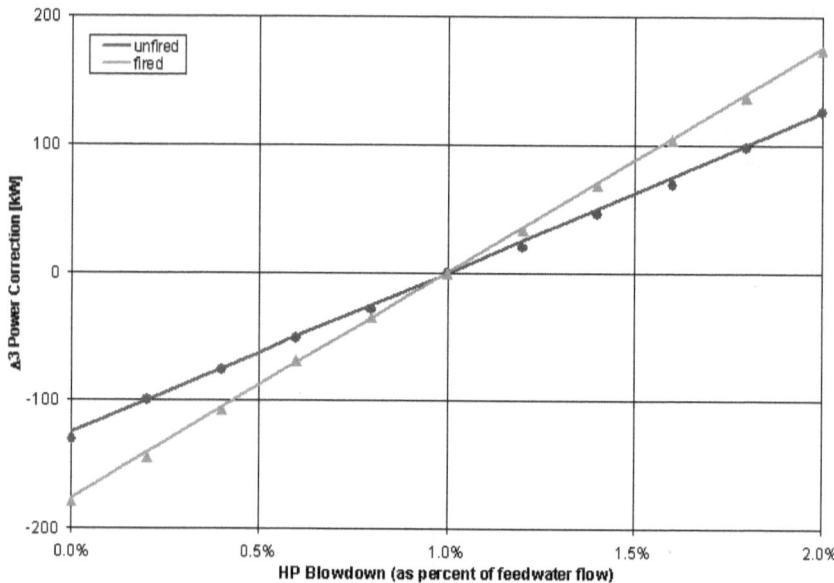

**Figure 6. Typical CCPP Delta3 Correction**

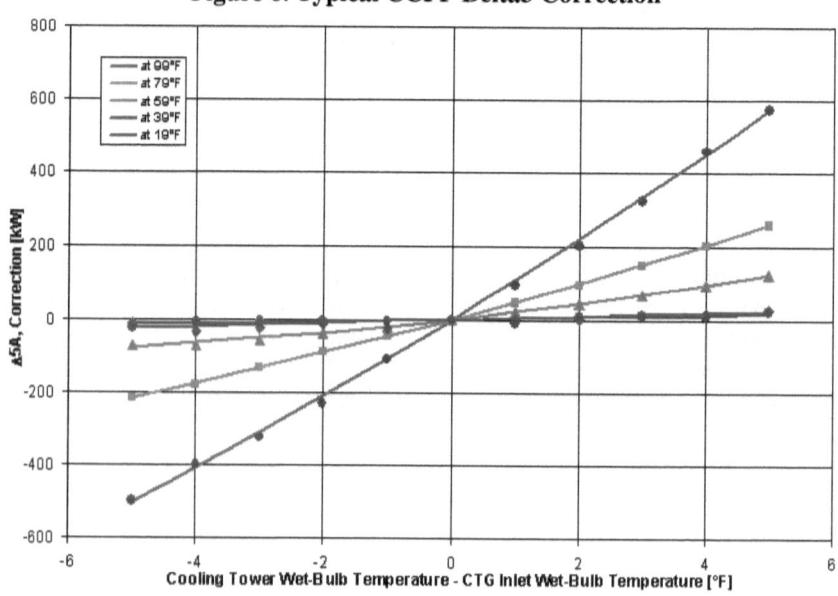

**Figure 7. Typical CCPP Delta5A Correction**

4

Note that both fired and unfired corrections are necessary when duct burners are present.

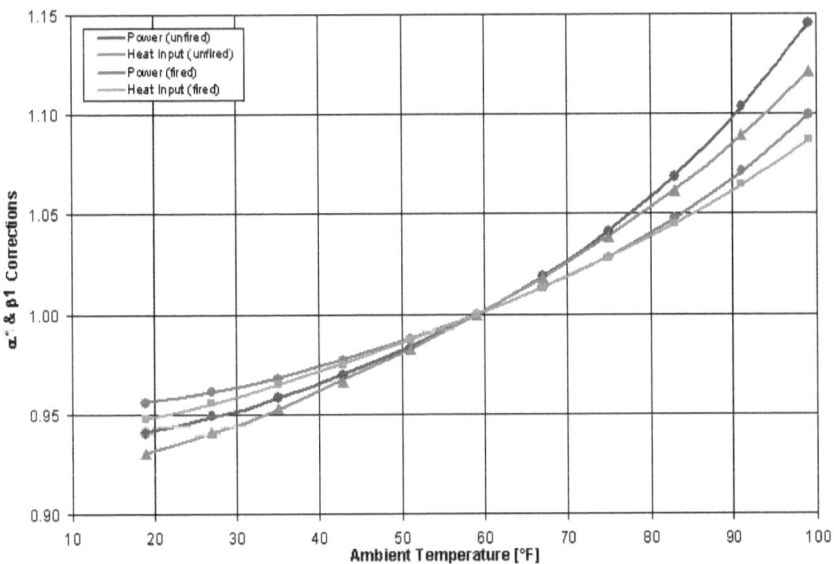

**Figure 8. Typical CCPP Alpha1 and Beta1 Corrections**

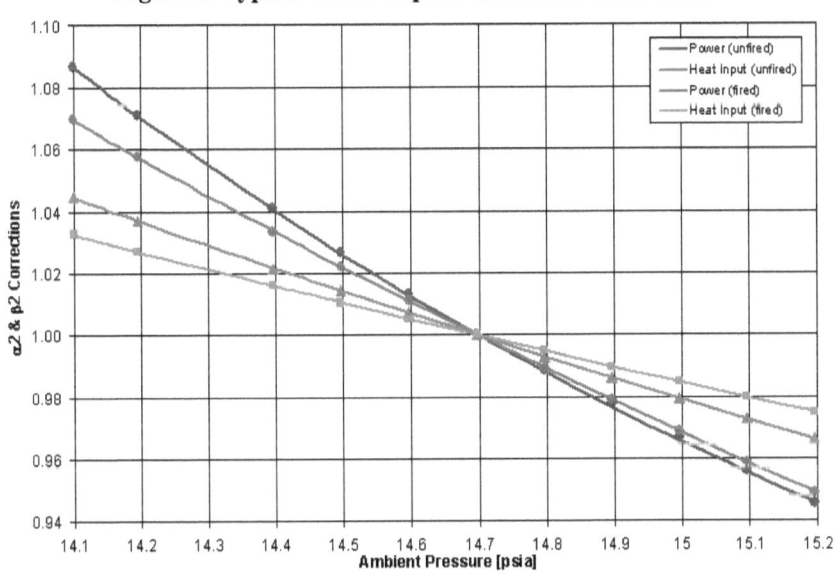

**Figure 9. Typical CCPP Alpha2 and Beta2 Corrections**

5

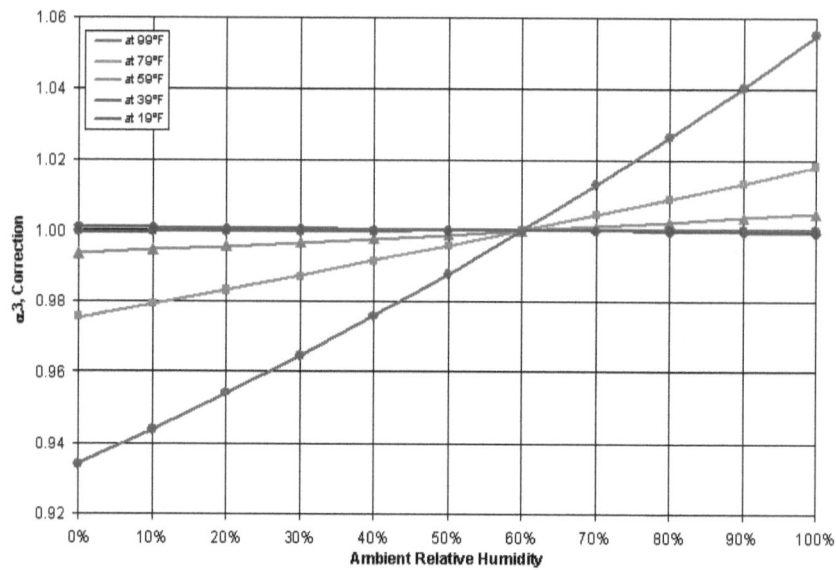

**Figure 10. Typical CCPP Alpha3 Correction**

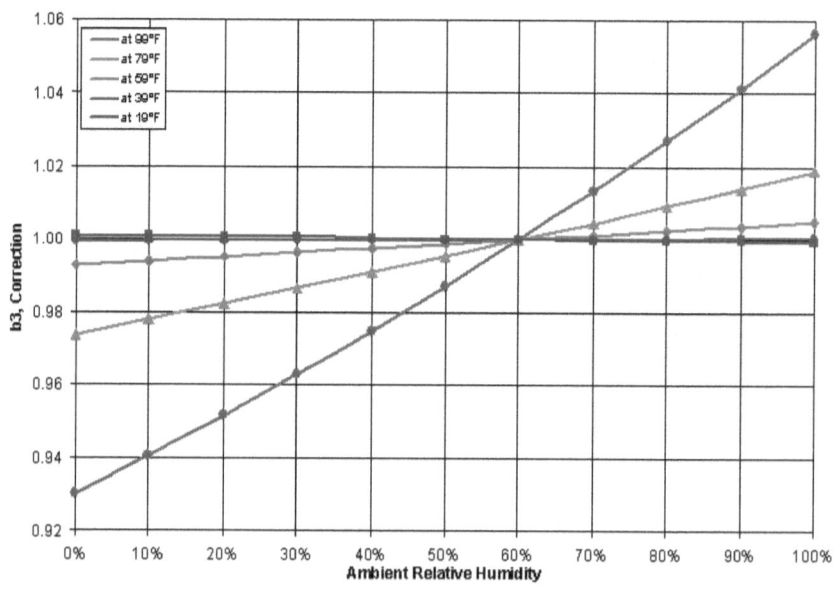

**Figure 11. Typical CCPP Beta3 Correction**

Relative humidity corrections may slope upward or downward, depending on the engines and system design. Fuel supply temperature is typically a very small correction, as it most often involves feedwater extraction.

6

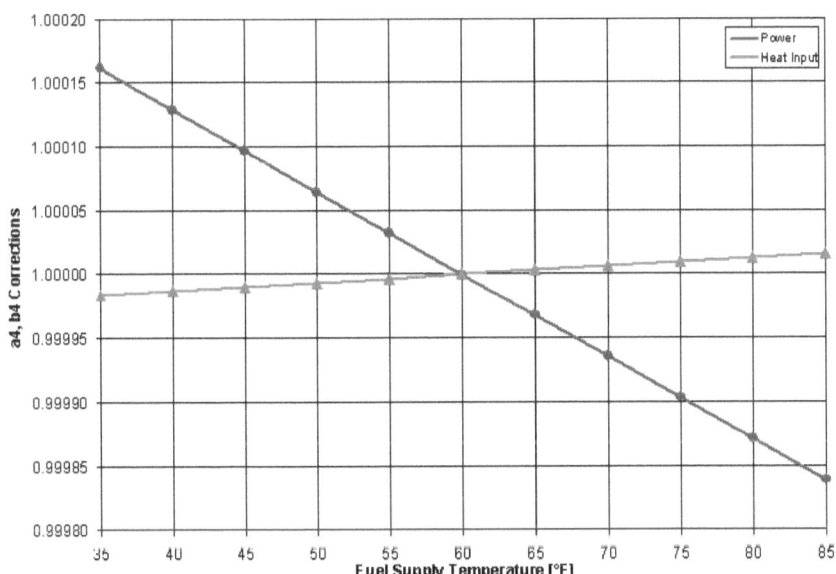

**Figure 12. Typical CCPP Alpha4 and Beta4 Corrections**

Fuel composition corrections, including carbon-to-hydrogen ratios and heating value are the source of much argument. Sometimes they slope upward and sometimes they slope downward. Sometimes the power and heat input corrections have opposite slopes. While there is some impact on the GT exhaust gas composition and thus specific heat and energy entering the HRSG, fuel composition also impacts emissions, whether directly or indirectly through settings and target values. The alpha5 and beta5 corrections should be small. If a system fails based on alpha5 or beta5, there is something else wrong somewhere.

The alpha5 and beta5 corrections are also not applied like the others. This is industry convention. These corrections are most often provided as three stacked straight lines. If there is any curvature to these corrections, it's most likely an error. The source of these corrections is not chemical in the sense of based on chemistry or thermodynamic calculations. They are empirical and deduced by the various engine manufacturers from testing experience. There simply isn't enough data (holding everything else constant) to infer anything beyond a bilinear relationship, that is, $z=a+bx+cy$; hence, the three stacked sloping straight lines. The corrections are applied using two linear adjustments: first for heating value and second for carbon-to-hydrogen ratio.

These corrections are typically supplied in terms of lower heating value (LHV), as higher heating value (HHV) is a meaningless quantity. People who insist on using HHV simply don't understand thermodynamics.

7

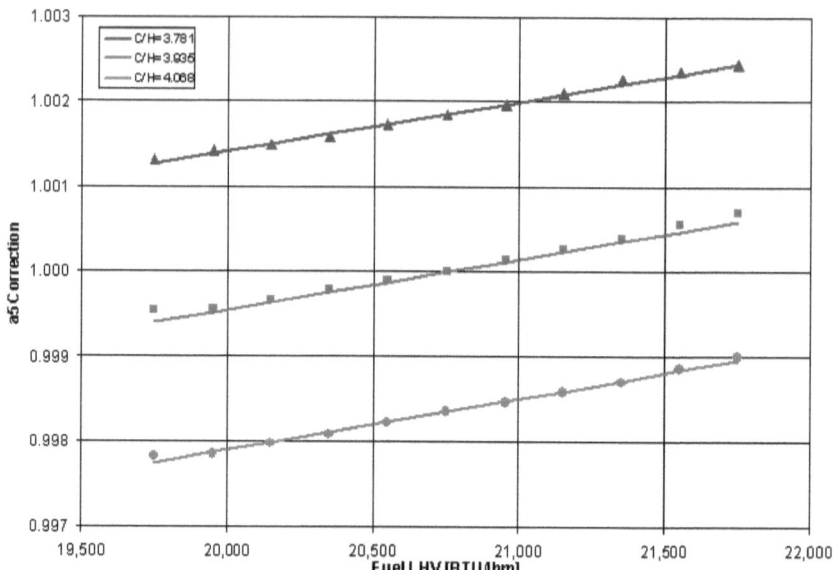

**Figure 13. Typical CCPP Alpha5 Correction**

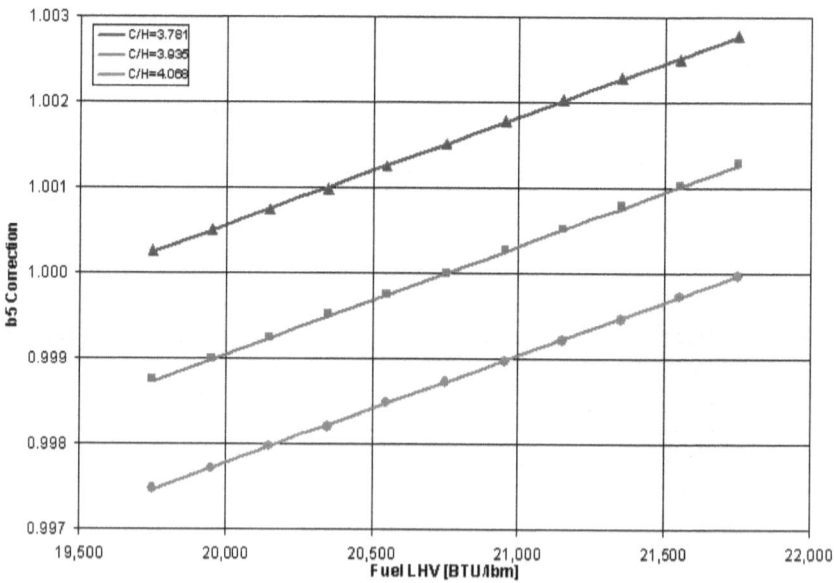

**Figure 14. Typical CCPP Beta5 Correction**

There may be corrections for age in terms of fired hours. Events (such as startup, shutdown, and cleaning) are accumulated in terms of equivalent hours.

8

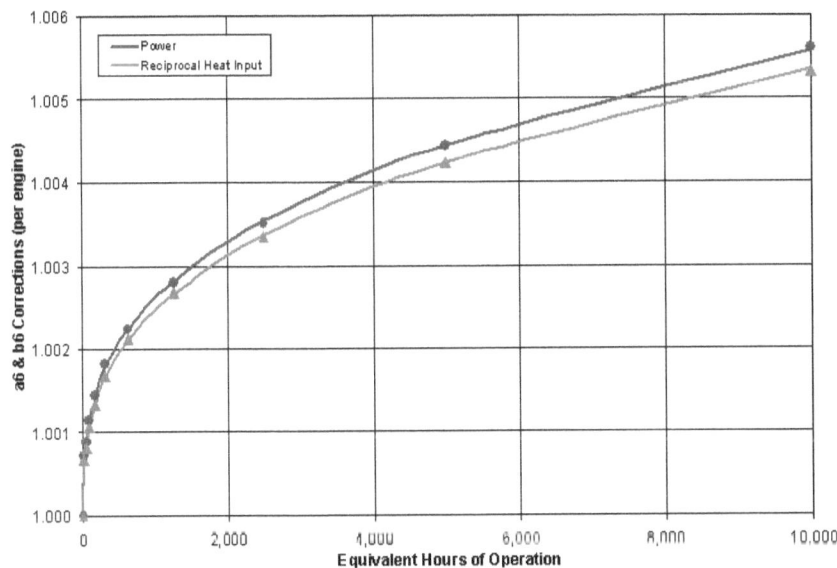

**Figure 15. Typical CCPP Alpha6 and Beta6 Corrections**

These corrections often are flat (no correction) out to some agreed upon interval and then rise sharply, as shown in the figure above. The impact is very steep at first and much less steep later in time, which requires progressively-spaced runs in order to generate a smooth curve. Note also, for convenience and comparison, the reciprocal ($1/y$) of the heat input correction is often plotted so that the correction must be applied accordingly.

### Duct Firing Curve

The last correction we will consider is the duct firing curve or Delta7-Omega7. This should be a downward-sloping straight line with a slope approximately equal to the heat rate. I you *over*-fire the duct burner during a test, then you should adjust the net power output downward accordingly; hence, the downward slope. If you *under*-fire during a test, then you should correct the net power output upward. For duct firing, the relationship between a change heat input and the corresponding change in net power output is fairly linear. If the correction isn't linear or doesn't have the expected slope, there is something wrong with the calculations.

You might wonder at this point how under-firing or over-firing during a test might be possible when you're trying to operate as close to design as possible. If you send gas samples to a laboratory for accurate testing, you may not know what the heating value is until days or week later; so you don't know exactly what the duct burner heat input is during a test. That's why you need a correction.

9

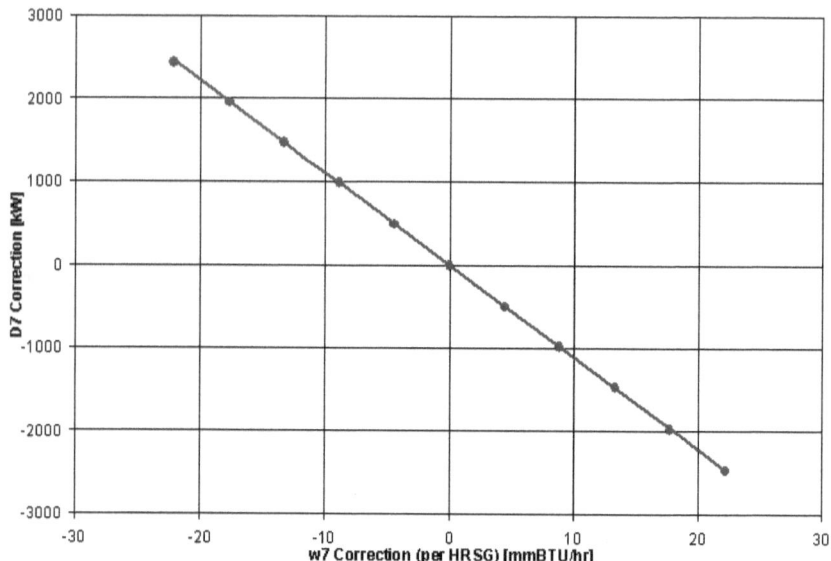

**Figure 16. Typical Delta7/Omega7 Duct Firing Correction**

We must also correct for power factor, as the plant must supply power to some grid and the power factor is considered outside control of the plant. We most often begin with generator efficiency curves (shown on the next page). From these we calculate losses by the following relationship. Sometimes loss curves are provided and this same relationship can be used to calculate efficiency.

$$efficiency = \frac{net\ power}{net\ power + losses} \quad (1.7)$$

The difference in generator losses at the as-tested power factor vs. design power factor at the as-tested net power yields the correction. If the losses are greater at the as-tested conditions than at the design conditions, then the difference is added to the as-tested power output (upward adjustment). If the losses are less at the as-tested conditions than at the design conditions, then the as-tested power output is adjusted downward. This is typically a small adjustment.

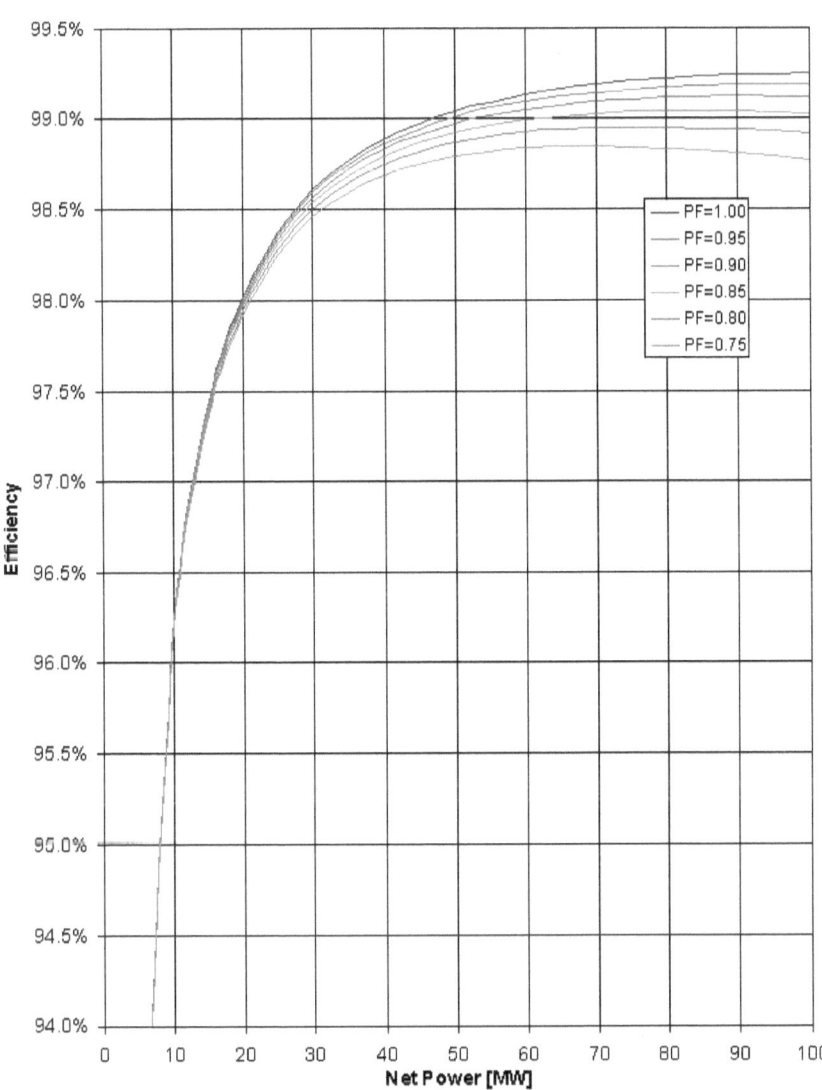

**Figure 17. Typical Generator Efficiency Curves**

Equation 1.7 relates these two sets of curves…

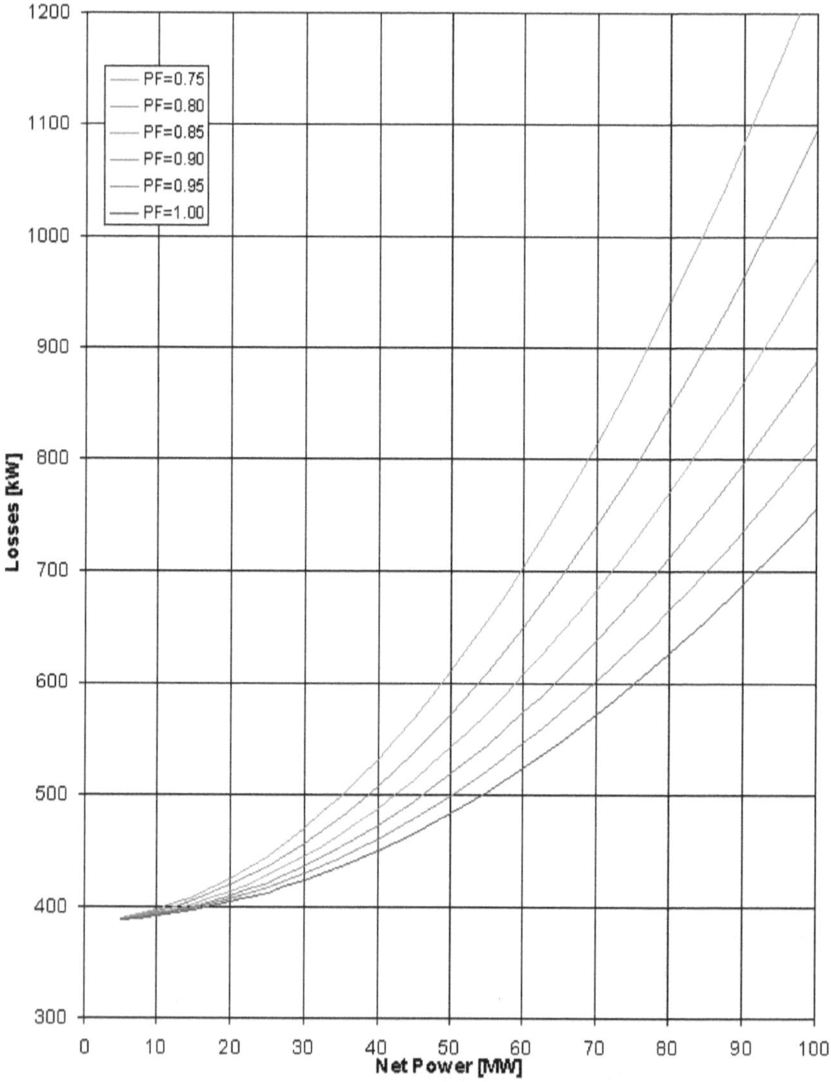

**Figure 18. Typical Generator Loss Curves**

The power factor correction is the difference in losses between two power factors and may be plotted as a function of power factor or net power.

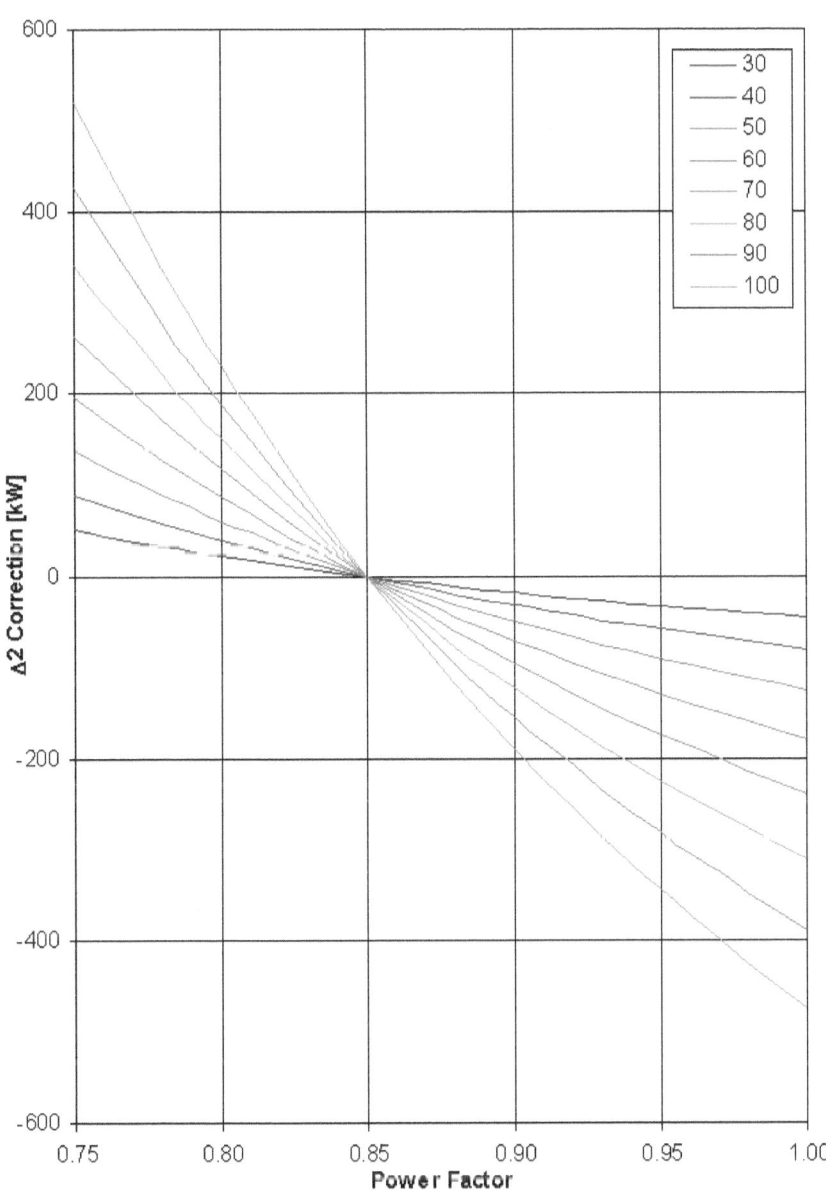

**Figure 19. Typical Type 1 Power Factor Correction**

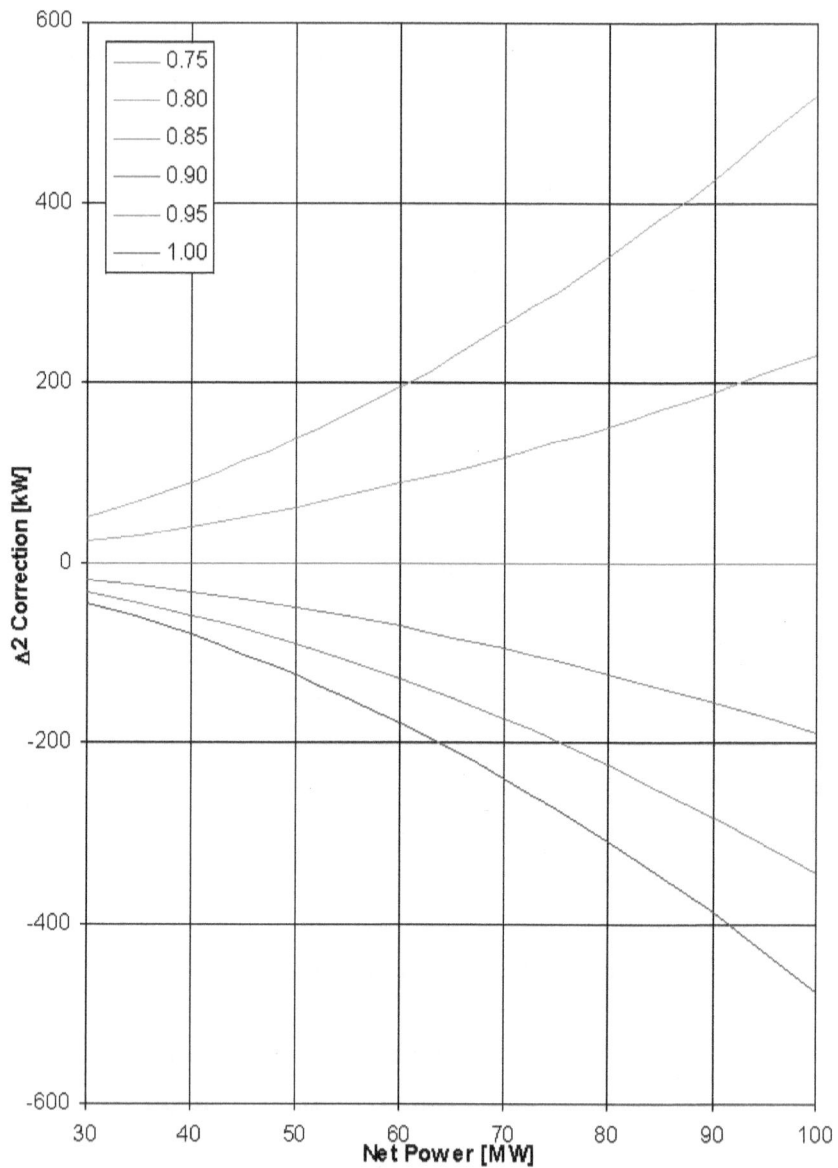

**Figure 20. Typical Type 2 Power Factor Correction**

## Applying the Corrections to a Test

We now step through the process of how these corrections are applied. The logical way to accomplish this is with an Excel spreadsheet, which is how we will do it here. All of the files (model, corrections, calculations, etc.) can be found in the online archive in folder examples. The model runs from which the corrections are derived are on the *unfired* and *fired* tabs, respectively. The coefficients for each curve are calculated just below the graph and linked to a comprehensive table on the *coefficients* tab. The test data, calculations, corrections, and results are on the *testing* tab. Model runs corresponding to each of the unfired and fired tests can be found on the *utest* and *ftest* tabs, respectively. It is customary to average four hour-long tests to arrive at a final result and so the spreadsheet (2x1CC_plant_curves.xls) is organized accordingly. The table of coefficients for unfired corrections appears on page iii and for fired corrections appears on page v. The input section appears below:

| TYPICAL 2X1 COMBINED CYCLE POWER PLANT CORRECTIONS FOR UNFIRED TESTING | | | | | | |
|---|---|---|---|---|---|---|
| | | UNFIRED | | | | |
| TEST CONDITIONS | units | Test1 | Test2 | Test3 | Test4 | Average |
| Barometric Pressure | psia | 14.60 | 14.62 | 14.88 | 14.88 | 14.85 |
| Ambient Dry-Bulb Temperature | °F | 61.3 | 57.7 | 61.5 | 61.1 | 60.4 |
| Ambient Wet-Bulb Temperature | °F | 51.0 | 49.1 | 50.8 | 51.1 | 50.5 |
| Cooling Tower Wet-Bulb | °F | 50.5 | 49.3 | 50.3 | 51.2 | 50.3 |
| Fuel Supply Temperature | °F | 53.9 | 53.1 | 53.7 | 53.1 | 53.4 |
| Fuel Gas CH4 Mole Frac. | - | 94.23% | 94.12% | 94.19% | 94.17% | 94.18% |
| Fuel Gas Ethane Mole Frac. | - | 0.93% | 0.87% | 0.93% | 0.89% | 0.91% |
| Fuel Gas Propane Mole Frac. | - | 0.98% | 1.03% | 1.01% | 1.01% | 1.01% |
| Fuel Gas n-Butane Mole Frac. | - | 0.48% | 0.50% | 0.51% | 0.48% | 0.49% |
| Fuel Gas Iso-Butane Mole Frac. | - | 0.51% | 0.52% | 0.50% | 0.52% | 0.51% |
| Fuel Gas n-Pentane Mole Frac. | - | 0.48% | 0.49% | 0.47% | 0.49% | 0.48% |
| Fuel Gas Iso-Pentane Mole Frac. | - | 0.53% | 0.53% | 0.51% | 0.50% | 0.52% |
| Fuel Gas n-Hexane(gas) Mole Frac. | - | 0.46% | 0.50% | 0.49% | 0.50% | 0.49% |
| Fuel Gas H2 Mole Frac. | - | 0.45% | 0.48% | 0.45% | 0.47% | 0.46% |
| Fuel Gas N2 Mole Frac. | - | 0.48% | 0.48% | 0.48% | 0.49% | 0.48% |
| Fuel Gas CO2 Mole Frac. | - | 0.45% | 0.49% | 0.47% | 0.48% | 0.47% |
| GT1 Equivaent Hours of Operation | EOH | 2728 | 2729 | 2730 | 2731 | 2730 |
| GT2 Equivaent Hours of Operation | EOH | 2621 | 2622 | 2623 | 2624 | 2623 |
| Power Factor | - | 0.891 | 0.892 | 0.893 | 0.891 | 0.892 |
| GT1 output | kW | 85,818 | 86,719 | 86,133 | 86,164 | 88,896 |
| GT2 output | kW | 85,521 | 86,569 | 85,984 | 86,015 | 89,038 |
| ST output | kW | 76,539 | 76,596 | 76,855 | 76,800 | 78,803 |
| Auxiliary Load | kW | 5,561 | 5,583 | 5,573 | 5,577 | 5,574 |
| Fuel Flow to GT1 | lbm/hr | 42,175 | 42,637 | 42,415 | 42,448 | 42,419 |
| Fuel Flow to GT2 | lbm/hr | 42,321 | 42,785 | 42,562 | 42,595 | 42,566 |
| Fuel Flow to DB1 | lbm/hr | 0 | 0 | 0 | 0 | 0 |
| Fuel Flow to DB2 | lbm/hr | 0 | 0 | 0 | 0 | 0 |
| HP Steam Flow A | lbm/hr | 283,361 | 283,035 | 283,808 | 283,885 | 283,422 |
| HP Steam Flow B | lbm/hr | 283,466 | 283,026 | 283,801 | 283,001 | 283,571 |
| HP Blowdown A | lbm/hr | 3,112 | 3,039 | 2,996 | 2,864 | 3,003 |
| HP Blowdown B | lbm/hr | 2,930 | 2,925 | 2,911 | 2,977 | 2,936 |

**Figure 21. Unfired Testing Inputs**

The calculations and power corrections are:

| CALCULATIONS | units | Test1 | Test2 | Test3 | Test4 | Average |
|---|---|---|---|---|---|---|
| Relative Humidity | - | 49% | 54% | 47% | 50% | 50% |
| CT Delta Wet-Bulb | °F | -0.5 | 0.2 | -0.6 | 0.1 | -0.2 |
| HP1 Blowdown fraction | - | 1.09% | 1.06% | 1.05% | 1.00% | 1.05% |
| HP2 Blowdown fraction | - | 1.02% | 1.02% | 1.01% | 1.04% | 1.02% |
| Net Power Output | kW | 242,317 | 244,302 | 243,398 | 243,402 | 243,355 |
| Fuel Gas Heating Value | BTU/lbm | 20,878 | 20,857 | 20,871 | 20,860 | 20,866 |
| Fuel Gas C/H Ratio | - | 3.805 | 3.803 | 3.805 | 3.805 | 3.804 |
| GT1 Heat Input | MBTU/hr | 880.5 | 889.3 | 885.2 | 885.4 | 885.1 |
| GT2 Heat Input | MBTU/hr | 883.6 | 892.4 | 888.3 | 888.5 | 888.2 |
| DB1 Heat Input | MBTU/hr | 0.0 | 0.0 | 0.0 | 0.0 | 0.0 |
| DB2 Heat Input | MBTU/hr | 0.0 | 0.0 | 0.0 | 0.0 | 0.0 |
| Total Heat Input | MBTU/hr | 1764.1 | 1781.7 | 1773.5 | 1774.0 | 1773.3 |
| **ADDITIVE POWER CORRECTIONS** | units | Test1 | Test2 | Test3 | Test4 | Average |
| Power Factor | kW | -330 | -342 | -347 | -332 | -338 |
| HP Blowdown | kW | 11 | 8 | 6 | 0 | 6 |
| CT Wet-Bulb at 99°F | 99 | 0 | 0 | 0 | 0 | 0 |
| CT Wet-Bulb at 79°F | 79 | 0 | 0 | 0 | 0 | 0 |
| CT Wet-Bulb at 59°F | 59 | 0 | 0 | 0 | 0 | 0 |
| CT Wet-Bulb at 39°F | 39 | 1 | 0 | 1 | 0 | 0 |
| CT Wet-Bulb at 19°F | 19 | 0 | 0 | 0 | 0 | 0 |
| CT Wet-Bulb at Test Conditions | kW | 0 | 0 | 0 | 0 | 0 |
| Duct Firing | kW | 0 | 0 | 0 | 0 | 0 |
| Total Additive Power Correction | kW | -318 | -334 | -341 | -332 | -332 |
| **MULTIPLICATIVE POWER CORRECTIONS** | units | Test1 | Test2 | Test3 | Test4 | Average |
| Ambient Dry-Bulb Temperature | - | 1.0056 | 0.9969 | 1.0060 | 1.0051 | 1.0034 |
| Barometric Pressure | - | 1.0064 | 1.0055 | 1.0011 | 1.0014 | 1.0036 |
| Relative Humidity at 99°F | 99 | 0.9881 | 0.9935 | 0.9867 | 0.9890 | 0.9893 |
| Relative Humidity at 79°F | 79 | 0.9958 | 0.9977 | 0.9953 | 0.9961 | 0.9962 |
| Relative Humidity at 59°F | 59 | 0.9989 | 0.9994 | 0.9988 | 0.9990 | 0.9990 |
| Relative Humidity at 39°F | 39 | 0.9999 | 1.0000 | 0.9999 | 0.9999 | 0.9999 |
| Relative Humidity at 19°F | 19 | 1.0002 | 1.0001 | 1.0002 | 1.0002 | 1.0002 |
| Relative Humidity at Test Conditions | - | 0.9985 | 0.9994 | 0.9983 | 0.9987 | 0.9987 |
| Fuel Supply Temperature | - | 1.0000 | 1.0000 | 1.0000 | 1.0000 | 1.0000 |
| Fuel Gas LHV at C/H=4.068 | 4.068 | 0.9983 | 0.9983 | 0.9983 | 0.9983 | 0.9983 |
| Fuel Gas LHV at C/H=3.935 | 3.935 | 1.0000 | 1.0000 | 1.0000 | 1.0000 | 1.0000 |
| Fuel Gas LHV at C/H=3.781 | 3.781 | 1.0018 | 1.0018 | 1.0018 | 1.0018 | 1.0018 |
| Fuel Gas LHV and C/C at Test Conditions | - | 1.0015 | 1.0015 | 1.0015 | 1.0015 | 1.0015 |
| Equivalent Hours of Operation | - | 1.0072 | 1.0072 | 1.0072 | 1.0072 | 1.0072 |
| Total Multiplicative Power Correction | - | 1.0194 | 1.0106 | 1.0142 | 1.0141 | 1.0146 |

**Figure 22. Unfired Calculations and Power Corrections**

The size of variations and deviations shown here are typical, as are the magnitude of the corrections. These should be no more than 10%. If any of the corrections is larger than 10%, this does not necessarily indicate a problem with the system, but it is preferable to test at conditions close enough to the guarantee point that the corrections are just that.

Note the numbers in the units column (e.g., 99, 79, 59, 39, 19) facilitate interpolation between the bivariate curves, which is a two-step process.

The heat input corrections and overall test results for the unfired cases are shown in this last section.

| ADDITIVE HEAT INPUT CORRECTION | units | Test1 | Test2 | Test3 | Test4 | Average |
|---|---|---|---|---|---|---|
| Duct Firing | MBTU/hr | 0.0 | 0.0 | 0.0 | 0.0 | 0.0 |
| **MULTIPLICATIVE HEAT INPUT CORRECTIONS** | units | Test1 | Test2 | Test3 | Test4 | Average |
| Ambient Dry-Bulb Temperature | - | 1.0065 | 0.9965 | 1.0069 | 1.0059 | 1.0039 |
| Barometric Pressure | - | 1.0067 | 1.0057 | 1.0012 | 1.0015 | 1.0038 |
| Relative Humidity at 99°F | 99 | 0.9856 | 0.9921 | 0.9839 | 0.9867 | 0.9871 |
| Relative Humidity at 79°F | 79 | 0.9949 | 0.9972 | 0.9943 | 0.9953 | 0.9954 |
| Relative Humidity at 59°F | 59 | 0.9986 | 0.9993 | 0.9985 | 0.9987 | 0.9988 |
| Relative Humidity at 39°F | 39 | 0.9999 | 0.9999 | 0.9999 | 0.9999 | 0.9999 |
| Relative Humidity at 19°F | 19 | 1.0002 | 1.0001 | 1.0002 | 1.0002 | 1.0002 |
| Relative Humidity at Test Conditions | - | 0.9982 | 0.9993 | 0.9979 | 0.9984 | 0.9985 |
| Fuel Supply Temperature | - | 1.0000 | 1.0000 | 1.0000 | 1.0000 | 1.0000 |
| Fuel Gas LHV at C/H=4.068 | 4.068 | 0.9987 | 0.9987 | 0.9987 | 0.9987 | 0.9987 |
| Fuel Gas LHV at C/H=3.935 | 3.935 | 1.0000 | 0.9999 | 0.9999 | 0.9999 | 0.9999 |
| Fuel Gas LHV at C/H=3.781 | 3.781 | 1.0014 | 1.0014 | 1.0014 | 1.0014 | 1.0014 |
| Fuel Gas LHV and C/H at Test Conditions | - | 1.0012 | 1.0012 | 1.0012 | 1.0012 | 1.0012 |
| Equivalent Hours of Operation (reciprocal) | - | 0.9965 | 0.9965 | 0.9965 | 0.9965 | 0.9965 |
| Total Multiplicative Heat Input Correction | - | 1.0091 | 0.9992 | 1.0037 | 1.0035 | 1.0039 |
| **CORRECTED TEST VALUES** | units | Test1 | Test2 | Test3 | Test4 | Average |
| Net Power | kW | 246,700 | 246,555 | 246,516 | 246,485 | 246,564 |
| Heat Input | MBTU/hr | 1780.1 | 1780.3 | 1780.1 | 1780.1 | 1780.2 |
| Heat Rate | BTU/kWh | 7215.7 | 7220.6 | 7220.9 | 7222.1 | 7219.8 |
| **CORRECTED TEST MARGINS** | units | Test1 | Test2 | Test3 | Test4 | Average |
| Net Power | - | 0.29% | 0.23% | 0.22% | 0.21% | 0.24% |
| Heat Input | - | -0.17% | -0.16% | -0.18% | -0.17% | -0.17% |
| Heat Rate | - | -0.46% | -0.40% | -0.39% | -0.38% | -0.41% |
| **CORRECTED TEST RESULTS** | units | Test1 | Test2 | Test3 | Test4 | Average |
| Net Power | - | PASS | PASS | PASS | PASS | PASS |
| Heat Input | - | PASS | PASS | PASS | PASS | PASS |
| Heat Rate | - | PASS | PASS | PASS | PASS | PASS |

**Figure 23. Unfired Heat Input Corrections and Overall Results**

For power, *passing* means a positive margin (i.e., a corrected test result greater than or equal to the guarantee). For heat input or heat rate, *passing* means a negative margin (i.e., a corrected test result less than or equal to the guarantee).

It is customary to wait at least four hours before testing after a change in operation, such as activating duct burners. Often the two tests are performed on subsequent days. The fired testing inputs are:

| TYPICAL 2X1 COMBINED CYCLE POWER PLANT CORRECTIONS FOR FIRED TESTING | | | | | | |
|---|---|---|---|---|---|---|
| | | FIRED | | | | |
| TEST CONDITIONS | units | Test1 | Test2 | Test3 | Test4 | Average |
| Barometric Pressure | psia | 14.67 | 14.61 | 14.61 | 14.64 | 14.63 |
| Ambient Dry-Bulb Temperature | °F | 59.3 | 58.7 | 59.1 | 59.4 | 59.1 |
| Ambient Wet-Bulb Temperature | °F | 50.1 | 50.4 | 49.3 | 49.5 | 49.8 |
| Cooling Tower Wet-Bulb | °F | 50.0 | 51.2 | 49.4 | 49.7 | 50.1 |
| Fuel Supply Temperature | °F | 53.1 | 53.0 | 53.2 | 53.2 | 53.1 |
| Fuel Gas CH4 Mole Frac. | - | 94.25% | 94.29% | 94.20% | 94.28% | 94.26% |
| Fuel Gas Ethane Mole Frac. | - | 0.90% | 0.90% | 0.94% | 0.97% | 0.93% |
| Fuel Gas Propane Mole Frac. | - | 0.94% | 0.90% | 0.96% | 0.90% | 0.92% |
| Fuel Gas n-Butane Mole Frac. | - | 0.45% | 0.47% | 0.46% | 0.44% | 0.46% |
| Fuel Gas Iso-Butane Mole Frac. | - | 0.54% | 0.55% | 0.52% | 0.49% | 0.53% |
| Fuel Gas n-Pentane Mole Frac. | - | 0.50% | 0.47% | 0.49% | 0.51% | 0.49% |
| Fuel Gas Iso-Pentane Mole Frac. | - | 0.50% | 0.50% | 0.51% | 0.52% | 0.51% |
| Fuel Gas n-Hexane(gas) Mole Frac. | - | 0.47% | 0.46% | 0.48% | 0.50% | 0.48% |
| Fuel Gas H2 Mole Frac. | - | 0.50% | 0.52% | 0.49% | 0.46% | 0.49% |
| Fuel Gas N2 Mole Frac. | - | 0.46% | 0.44% | 0.48% | 0.49% | 0.47% |
| Fuel Gas CO2 Mole Frac. | - | 0.49% | 0.50% | 0.47% | 0.44% | 0.47% |
| GT1 Equivaent Hours of Operation | EOH | 2752 | 2753 | 2754 | 2755 | 2754 |
| GT2 Equivaent Hours of Operation | EOH | 2645 | 2646 | 2647 | 2648 | 2647 |
| Power Factor | - | 0.894 | 0.893 | 0.891 | 0.890 | 0.892 |
| GT1 output | kW | 87,106 | 86,776 | 86,803 | 86,842 | 86,882 |
| GT2 output | kW | 86,793 | 86,665 | 86,622 | 86,727 | 86,702 |
| ST output | kW | 96,247 | 96,010 | 96,279 | 96,252 | 96,197 |
| Auxiliary Load | kW | 6,664 | 6,695 | 6,668 | 6,674 | 6,675 |
| Fuel Flow to GT1 | lbm/hr | 42,447 | 42,334 | 42,338 | 42,354 | 42,368 |
| Fuel Flow to GT2 | lbm/hr | 42,596 | 42,480 | 42,469 | 42,475 | 42,505 |
| Fuel Flow to DB1 | lbm/hr | 4,249 | 4,266 | 4,320 | 4,274 | 4,277 |
| Fuel Flow to DB2 | lbm/hr | 4,263 | 4,277 | 4,329 | 4,285 | 4,288 |
| HP Steam Flow A | lbm/hr | 367,342 | 367,401 | 367,204 | 367,265 | 367,303 |
| HP Steam Flow B | lbm/hr | 367,420 | 367,225 | 367,204 | 367,760 | 367,402 |
| HP Blowdown A | lbm/hr | 3,977 | 3,963 | 3,715 | 3,637 | 3,823 |
| HP Blowdown B | lbm/hr | 3,532 | 3,986 | 3,774 | 3,665 | 3,739 |

**Figure 24. Fired Testing Inputs**

The engines (combustion gas turbines) should operate quite similar whether duct firing is activated or not. The only substantial difference might be a small increase in pressure drop across the HRSG due to the increase volumetric flow, resulting from additional fuel and higher temperature. If it is not, this is a problem that should be investigated further. It is always advisable to perform a turbine test (PTC-22) before testing the entire combined cycle.[5]

---

[5] "PTC-22 Gas Turbines," American Society of Mechanical Engineers, 2005.

The calculation and power correction section follows:

| CALCULATIONS | units | Test1 | Test2 | Test3 | Test4 | Average |
|---|---|---|---|---|---|---|
| Relative Humidity | - | 52% | 56% | 49% | 49% | 52% |
| CT Delta Wet-Bulb | °F | -0.1 | 0.8 | 0.2 | 0.2 | 0.3 |
| HP1 Blowdown fraction | - | 1.07% | 1.07% | 1.00% | 0.98% | 1.03% |
| HP2 Blowdown fraction | - | 0.95% | 1.07% | 1.02% | 0.99% | 1.01% |
| Net Power Output | kW | 263,482 | 262,756 | 263,036 | 263,147 | 263,105 |
| Fuel Gas Heating Value | BTU/lbm | 20,870 | 20,871 | 20,872 | 20,885 | 20,874 |
| Fuel Gas C/H Ratio | - | 3.808 | 3.810 | 3.806 | 3.804 | 3.807 |
| GT1 Heat Input | MBTU/hr | 885.9 | 883.5 | 883.7 | 884.6 | 884.4 |
| GT2 Heat Input | MBTU/hr | 889.0 | 886.6 | 886.4 | 887.1 | 887.3 |
| DB1 Heat Input | MBTU/hr | 88.7 | 89.0 | 90.2 | 89.3 | 89.3 |
| DB2 Heat Input | MBTU/hr | 89.0 | 89.3 | 90.4 | 89.5 | 89.5 |
| Total Heat Input | MBTU/hr | 1952.5 | 1948.4 | 1950.6 | 1950.4 | 1950.5 |
| **ADDITIVE POWER CORRECTIONS** | units | Test1 | Test2 | Test3 | Test4 | Average |
| Power Factor | kW | -414 | -404 | -387 | -379 | -396 |
| HP Blowdown | kW | 12 | 12 | 0 | -3 | 5 |
| CT Wet-Bulb at 99°F | 99 | -14 | 109 | 23 | 25 | 36 |
| CT Wet-Bulb at 79°F | 79 | -7 | 51 | 11 | 11 | 17 |
| CT Wet-Bulb at 59°F | 59 | -4 | 27 | 6 | 6 | 9 |
| CT Wet-Bulb at 39°F | 39 | -1 | 6 | 1 | 1 | 2 |
| CT Wet-Bulb at 19°F | 19 | -1 | 4 | 1 | 1 | 1 |
| CT Wet-Bulb at Test Conditions | kW | -4 | 27 | 6 | 6 | 9 |
| Duct Firing | kW | -29 | -100 | -351 | -151 | -158 |
| Total Additive Power Correction | kW | -434 | -466 | -732 | -527 | -540 |
| **MULTIPLICATIVE POWER CORRECTIONS** | units | Test1 | Test2 | Test3 | Test4 | Average |
| Ambient Dry-Bulb Temperature | - | 1.0007 | 0.9993 | 1.0002 | 1.0009 | 1.0003 |
| Barometric Pressure | - | 1.0020 | 1.0053 | 1.0054 | 1.0039 | 1.0041 |
| Relative Humidity at 99°F | 99 | 0.9907 | 0.9954 | 0.9870 | 0.9871 | 0.9900 |
| Relative Humidity at 79°F | 79 | 0.9967 | 0.9984 | 0.9954 | 0.9954 | 0.9965 |
| Relative Humidity at 59°F | 59 | 0.9991 | 0.9996 | 0.9987 | 0.9987 | 0.9990 |
| Relative Humidity at 39°F | 39 | 0.9999 | 1.0000 | 0.9999 | 0.9999 | 0.9999 |
| Relative Humidity at 19°F | 19 | 1.0001 | 1.0001 | 1.0002 | 1.0002 | 1.0001 |
| Relative Humidity at Test Conditions | - | 0.9991 | 0.9996 | 0.9987 | 0.9987 | 0.9990 |
| Fuel Supply Temperature | - | 1.0001 | 1.0001 | 1.0001 | 1.0001 | 1.0001 |
| Fuel Gas LHV at C/H=4.068 | 4.068 | 0.9985 | 0.9985 | 0.9985 | 0.9985 | 0.9985 |
| Fuel Gas LHV at C/H=3.935 | 3.935 | 1.0000 | 1.0000 | 1.0000 | 1.0000 | 1.0000 |
| Fuel Gas LHV at C/H=3.781 | 3.781 | 1.0017 | 1.0017 | 1.0017 | 1.0017 | 1.0017 |
| Fuel Gas LHV and C/C at Test Conditions | - | 1.0014 | 1.0014 | 1.0014 | 1.0014 | 1.0014 |
| Equivalent Hours of Operation | - | 1.0067 | 1.0067 | 1.0067 | 1.0067 | 1.0067 |
| Total Multiplicative Power Correction | - | 1.0099 | 1.0123 | 1.0126 | 1.0116 | 1.0116 |

**Figure 25. Fired Calculations and Power Corrections**

The heat input, heat rate, and overall test results are:

| ADDITIVE HEAT INPUT CORRECTION | units | Test1 | Test2 | Test3 | Test4 | Average |
|---|---|---|---|---|---|---|
| Duct Firing | MBTU/hr | 0.3 | 0.9 | 3.2 | 1.4 | 1.4 |
| **MULTIPLICATIVE HEAT INPUT CORRECTIONS** | units | Test1 | Test2 | Test3 | Test4 | Average |
| Ambient Dry-Bulb Temperature | - | 1.0008 | 0.9993 | 1.0003 | 1.0009 | 1.0003 |
| Barometric Pressure | - | 1.0020 | 1.0054 | 1.0055 | 1.0039 | 1.0042 |
| Relative Humidity at 99°F | 99 | 0.9910 | 0.9955 | 0.9874 | 0.9875 | 0.9903 |
| Relative Humidity at 79°F | 79 | 0.9968 | 0.9984 | 0.9955 | 0.9955 | 0.9966 |
| Relative Humidity at 59°F | 59 | 0.9991 | 0.9996 | 0.9988 | 0.9988 | 0.9991 |
| Relative Humidity at 39°F | 39 | 0.9999 | 1.0000 | 0.9999 | 0.9999 | 0.9999 |
| Relative Humidity at 19°F | 19 | 1.0001 | 1.0001 | 1.0002 | 1.0002 | 1.0001 |
| Relative Humidity at Test Conditions | - | 0.9991 | 0.9996 | 0.9988 | 0.9987 | 0.9991 |
| Fuel Supply Temperature | - | 1.0000 | 1.0000 | 1.0000 | 1.0000 | 1.0000 |
| Fuel Gas LHV at C/H=4.068 | 4.068 | 0.9988 | 0.9988 | 0.9988 | 0.9988 | 0.9988 |
| Fuel Gas LHV at C/H=3.935 | 3.935 | 0.9999 | 0.9999 | 1.0000 | 1.0000 | 1.0000 |
| Fuel Gas LHV at C/H=3.781 | 3.781 | 1.0013 | 1.0013 | 1.0013 | 1.0013 | 1.0013 |
| Fuel Gas LHV and C/H at Test Conditions | - | 1.0011 | 1.0011 | 1.0011 | 1.0011 | 1.0011 |
| Equivalent Hours of Operation (reciprocal) | - | 0.9968 | 0.9968 | 0.9968 | 0.9968 | 0.9968 |
| Total Multiplicative Heat Input Correction | - | 0.9998 | 1.0021 | 1.0024 | 1.0015 | 1.0015 |
| **CORRECTED TEST VALUES** | units | Test1 | Test2 | Test3 | Test4 | Average |
| Net Power | kW | 265,649 | 265,518 | 265,601 | 265,675 | 265,611 |
| Heat Input | MBTU/hr | 1952.3 | 1953.4 | 1958.5 | 1954.7 | 1954.7 |
| Heat Rate | BTU/kWh | 7349.0 | 7356.8 | 7374.0 | 7357.6 | 7359.3 |
| **CORRECTED TEST MARGINS** | units | Test1 | Test2 | Test3 | Test4 | Average |
| Net Power | - | 0.62% | 0.57% | 0.60% | 0.63% | 0.60% |
| Heat Input | - | -0.31% | -0.25% | 0.01% | -0.18% | -0.18% |
| Heat Rate | - | -0.92% | -0.81% | -0.58% | -0.80% | -0.78% |
| **CORRECTED TEST RESULTS** | units | Test1 | Test2 | Test3 | Test4 | Average |
| Net Power | - | PASS | PASS | PASS | PASS | PASS |
| Heat Input | - | PASS | PASS | FAIL | PASS | PASS |
| Heat Rate | - | PASS | PASS | PASS | PASS | PASS |

**Figure 26. Fired Heat Input Corrections and Overall Results**

It is not uncommon for one of the four test periods to fail by a slight amount. If the others pass adequately, this may be attributed to variability such as shifting weather, including wind direction, rising or falling barometric pressure. The example shown here is no reason for concern.

You may also want to run the model (in this case GateCycle™) at the test conditions for comparison. Model runs corresponding to the unfired test are listed below:

| | Units | Design | Test1 | Test2 | Test3 | Test4 | Average |
|---|---|---|---|---|---|---|---|
| **GateCycle Inputs** | | | | | | | |
| Ambient Pressure | psia | 14.70 | 14.60 | 14.62 | 14.68 | 14.68 | 14.65 |
| Ambient Temperature | F | 59.0 | 61.3 | 57.7 | 61.5 | 61.1 | 60.4 |
| Ambient Relative Humidity | | 60% | 49% | 54% | 47% | 50% | 50% |
| delta wet-bulb | | 0.0 | -0.5 | 0.2 | -0.6 | 0.1 | -0.2 |
| Outlet Temperature | F | 60.0 | 53.9 | 53.1 | 53.7 | 53.1 | 53.4 |
| Fuel Gas CH4 Mole Frac. | | 94.00% | 94.23% | 94.12% | 94.19% | 94.17% | 94.18% |
| Fuel Gas Ethane Mole Frac. | | 2.50% | 0.93% | 0.87% | 0.93% | 0.89% | 0.91% |
| Fuel Gas Propane Mole Frac. | | 2.50% | 0.98% | 1.03% | 1.01% | 1.01% | 1.01% |
| Fuel Gas n-Butane Mole Frac. | | 0.00% | 0.48% | 0.50% | 0.51% | 0.48% | 0.49% |
| Fuel Gas Iso-Butane Mole Frac. | | 0.00% | 0.51% | 0.52% | 0.50% | 0.52% | 0.51% |
| Fuel Gas n-Pentane Mole Frac. | | 0.00% | 0.48% | 0.49% | 0.47% | 0.49% | 0.48% |
| Fuel Gas Iso-Pentane Mole Frac. | | 0.00% | 0.53% | 0.53% | 0.51% | 0.50% | 0.52% |
| Fuel Gas n-Hexane(gas) Mole Frac. | | 0.00% | 0.48% | 0.50% | 0.49% | 0.50% | 0.49% |
| Fuel Gas H2 Mole Frac. | | 0.00% | 0.45% | 0.48% | 0.45% | 0.47% | 0.46% |
| Fuel Gas N2 Mole Frac. | | 0.50% | 0.48% | 0.48% | 0.49% | 0.49% | 0.49% |
| Fuel Gas CO2 Mole Frac. | | 0.50% | 0.45% | 0.49% | 0.47% | 0.48% | 0.47% |
| Fired Hours | | 0 | 2,675 | 2,676 | 2,677 | 2,678 | 2,676 |
| Desired Blowdown as BFW Fraction | | 1.00% | 1.05% | 1.04% | 1.03% | 1.02% | 1.04% |
| Power Factor | | 0.850 | 0.891 | 0.892 | 0.893 | 0.891 | 0.892 |
| Power Factor | | 0.850 | 0.891 | 0.892 | 0.893 | 0.891 | 0.892 |
| Power Factor | | 0.850 | 0.891 | 0.892 | 0.893 | 0.891 | 0.892 |
| **GateCycle Outputs** | | | | | | | |
| Fuel Gas LHV | BTU/lb | 20,915 | 20,878 | 20,857 | 20,871 | 20,860 | 20,866 |
| H/C ratio | | 3.935 | 3.805 | 3.803 | 3.805 | 3.805 | 3.804 |
| Net Power | MW | 87.15 | 85.38 | 86.43 | 85.85 | 85.88 | 85.88 |
| Generator Terminal Power | k*kW | 77.28 | 76.28 | 76.33 | 76.61 | 76.54 | 76.44 |
| Steam Cycle BOP Losses | k*kW | 5.60 | 5.58 | 5.58 | 5.58 | 5.58 | 5.58 |
| Net Cycle Power | MW | 245.98 | 241.47 | 243.61 | 242.71 | 242.70 | 242.62 |
| Primary Fuel Inlet Flow | lb/hr | 42,630 | 42,461 | 42,926 | 42,701 | 42,735 | 42,705 |
| Heat Consumption | MM BTU/hr | 891.60 | 886.49 | 895.32 | 891.19 | 891.44 | 891.09 |
| Total LHV Fuel Cons. | MM BTU/hr | 1783.2 | 1773.0 | 1790.6 | 1782.4 | 1782.9 | 1782.2 |
| Net Heat Rate | BTU/kW-hr | 10,231 | 10,383 | 10,359 | 10,381 | 10,381 | 10,376 |
| Net Cycle LHV Heat Rate | BTU/kW-hr | 7249.4 | 7342.6 | 7350.6 | 7343.6 | 7345.9 | 7345.7 |

**Figure 27. GateCycle™ Model Runs Corresponding to Unfired Test**

If the heat balance model run at the as-tested conditions yields lower net power output (higher heat input or heat rate) than the uncorrected test result, this would indicate *passing*. The opposite would indicate *failing*.

21

Model runs for the fired test conditions are:

| Description | Units | Design | Test1 | Test2 | Test3 | Test4 | Average |
|---|---|---|---|---|---|---|---|
| **GateCycle Inputs** | | | | | | | |
| Ambient Pressure | psia | 14.70 | 14.67 | 14.61 | 14.61 | 14.64 | 14.63 |
| Ambient Temperature | F | 59.0 | 59.3 | 58.7 | 59.1 | 59.4 | 59.1 |
| Ambient Relative Humidity | | 60% | 52% | 56% | 49% | 49% | 52% |
| delta wet-bulb | | 0.0 | -0.1 | 0.8 | 0.2 | 0.2 | 0.3 |
| Outlet Temperature | F | 60.0 | 53.1 | 53.0 | 53.2 | 53.2 | 53.1 |
| Fuel Gas CH4 Mole Frac. | | 94.00% | 94.25% | 94.29% | 94.20% | 94.28% | 94.26% |
| Fuel Gas Ethane Mole Frac. | | 2.50% | 0.90% | 0.90% | 0.94% | 0.97% | 0.93% |
| Fuel Gas Propane Mole Frac. | | 2.50% | 0.94% | 0.90% | 0.96% | 0.90% | 0.92% |
| Fuel Gas n-Butane Mole Frac. | | 0.00% | 0.45% | 0.47% | 0.46% | 0.44% | 0.46% |
| Fuel Gas Iso-Butane Mole Frac. | | 0.00% | 0.54% | 0.55% | 0.52% | 0.49% | 0.53% |
| Fuel Gas n-Pentane Mole Frac. | | 0.00% | 0.50% | 0.47% | 0.49% | 0.51% | 0.49% |
| Fuel Gas Iso-Pentane Mole Frac. | | 0.00% | 0.50% | 0.50% | 0.51% | 0.52% | 0.51% |
| Fuel Gas n-Hexane(gas) Mole Frac. | | 0.00% | 0.47% | 0.46% | 0.48% | 0.50% | 0.48% |
| Fuel Gas H2 Mole Frac. | | 0.00% | 0.50% | 0.52% | 0.49% | 0.46% | 0.49% |
| Fuel Gas N2 Mole Frac. | | 0.50% | 0.46% | 0.44% | 0.48% | 0.49% | 0.47% |
| Fuel Gas CO2 Mole Frac. | | 0.50% | 0.49% | 0.50% | 0.47% | 0.44% | 0.47% |
| Desired Fuel Mass Flow | lb/hr | 4,265 | 4,249 | 4,266 | 4,300 | 4,274 | 4,272 |
| Desired Fuel Mass Flow | lb/hr | 4,265 | 4,263 | 4,277 | 4,309 | 4,285 | 4,283 |
| Fired Hours | | 0 | 2,699 | 2,700 | 2,701 | 2,702 | 2,700 |
| Desired Blowdown as BFW Fraction | | 1.00% | 1.01% | 1.07% | 1.01% | 0.98% | 1.02% |
| Power Factor | | 0.850 | 0.894 | 0.893 | 0.891 | 0.890 | 0.892 |
| Power Factor | | 0.850 | 0.894 | 0.893 | 0.891 | 0.890 | 0.892 |
| Power Factor | | 0.850 | 0.894 | 0.893 | 0.891 | 0.890 | 0.892 |
| **GateCycle Outputs** | | | | | | | |
| Fuel Gas LHV | BTU/lb | 20,915 | 20,870 | 20,871 | 20,872 | 20,885 | 20,874 |
| H/C ratio | | 3.935 | 3.808 | 3.810 | 3.806 | 3.804 | 3.807 |
| Net Power | MW | 87.14 | 86.89 | 86.68 | 86.65 | 86.72 | 86.73 |
| Generator Terminal Power | k*kW | 96.42 | 96.18 | 95.98 | 96.17 | 96.16 | 96.12 |
| Steam Cycle BOP Losses | k*kW | 6.71 | 6.71 | 6.70 | 6.70 | 6.70 | 6.70 |
| Net Cycle Power | MW | 263.99 | 263.26 | 262.64 | 262.76 | 262.89 | 262.89 |
| Primary Fuel Inlet Flow | lb/hr | 42,626 | 42,581 | 42,471 | 42,450 | 42,467 | 42,491 |
| Heat Consumption | MM BTU/hr | 891.52 | 888.67 | 886.40 | 885.99 | 886.91 | 886.96 |
| Total LHV Fuel Cons. | MM BTU/hr | 1961.4 | 1955.0 | 1951.1 | 1951.7 | 1952.5 | 1952.5 |
| Net Heat Rate | BTU/kW-hr | 10,231 | 10,227 | 10,226 | 10,225 | 10,227 | 10,226 |
| Net Cycle LHV Heat Rate | BTU/kW-hr | 7430.0 | 7426.2 | 7428.6 | 7427.6 | 7427.1 | 7427.2 |

**Figure 28. GateCycle™ Model Runs Corresponding to Fired Test**

## Chapter 2. Gas Turbine Performance Curves

Gas turbine manufacturers most often provide *performance* curves, not *correction* curves. These indicate how the engine will perform, not how to correct measured performance back to the design or reference conditions. If there is any doubt, check the ambient temperature or barometric pressure curves. Higher ambient temperatures will result in lower power output and this curve will drop more steeply as the temperature increases. Power output of a gas turbine is roughly proportional to barometric pressure, so these should be upward sloping and fairly linear.

GT curves most often include three multiplicative (power, heat rate, and exhaust flow) and one additive (exhaust temperature). These four are often abbreviated: POW, HRT, EGW, and EGT, respectively. Compressor inlet temperature is often abbreviated: CIT. The first three are shown on the left vertical scale and the fourth on the right vertical scale in the following figure.

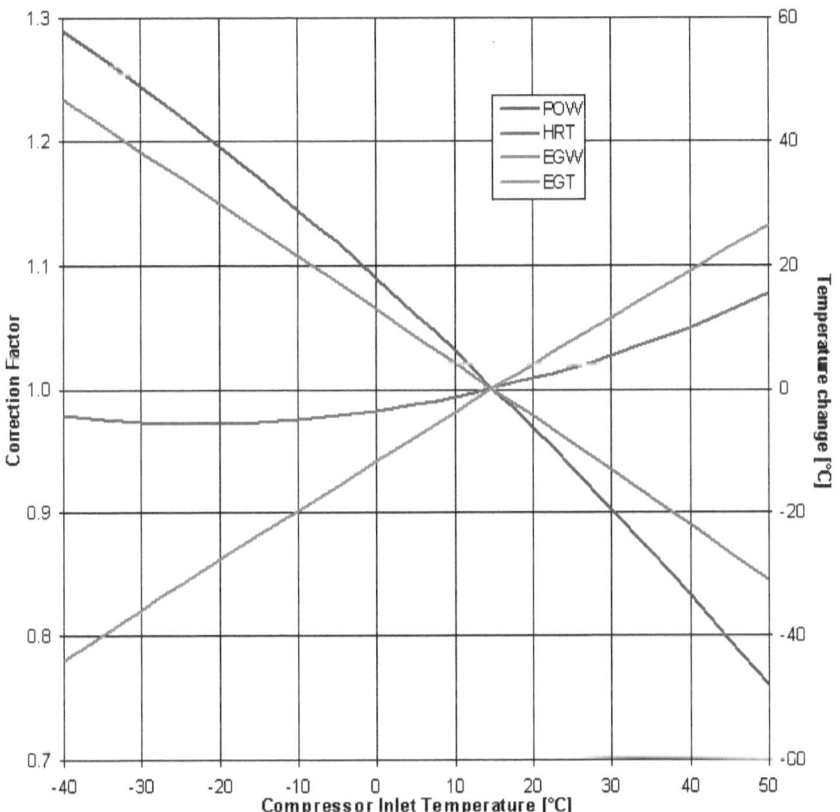

**Figure 29. Typical Gas Turbine Curves for Inlet Temperature**

23

There is often some curvature in all of these except for the fourth, EGT. Curves vary from engine-to-engine with make and model as well as local tuning and emission controls, which can depend on the fuel. Note that all of the multiplicative curves must pass through one and the additive curves through zero at the reference (or base conditions).

The impact of barometric pressure is typically linear, as shown in the next figure. Power and fuel consumption are almost linear with pressure, making heat rate vary only slightly with barometric pressure. The change in exhaust temperature is usually linear. Curves which exhibit an abrupt change in slope are indicative of an operational change, such as implemented by the control system to meet emissions requirements or perhaps a shaft limit. Shaft limits are quite problematic and you should refuse to purchase an engine with a shaft limit.

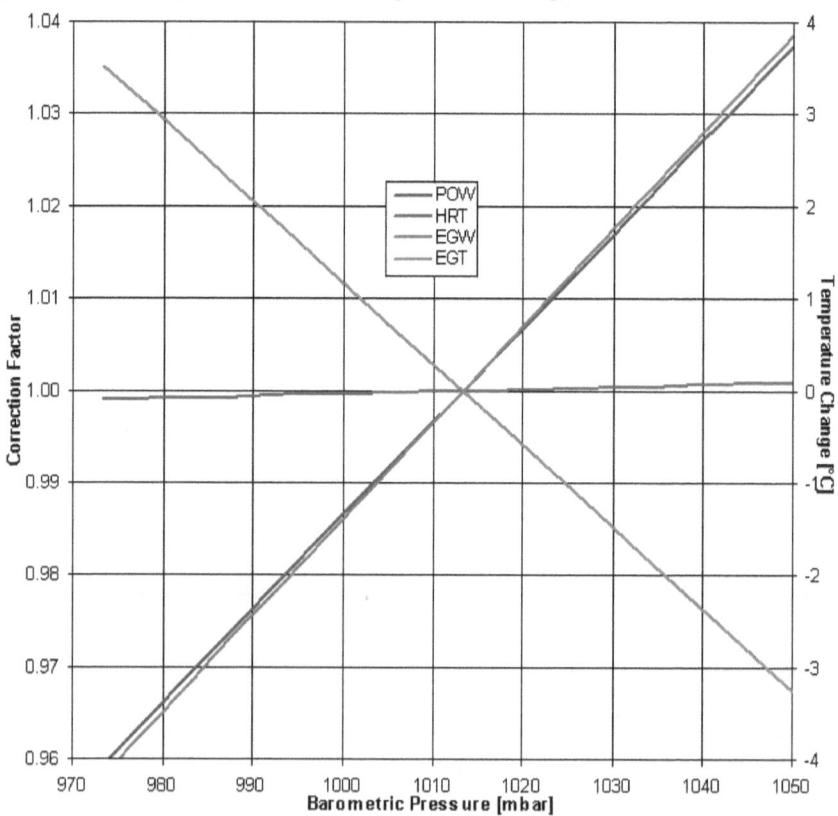

**Figure 30. Typical Gas Turbine Curves for Barometric Pressure**

The impact of relative humidity also depends on ambient temperature, which necessitates a fan of curves for each correction, as illustrated in the next four graphs.

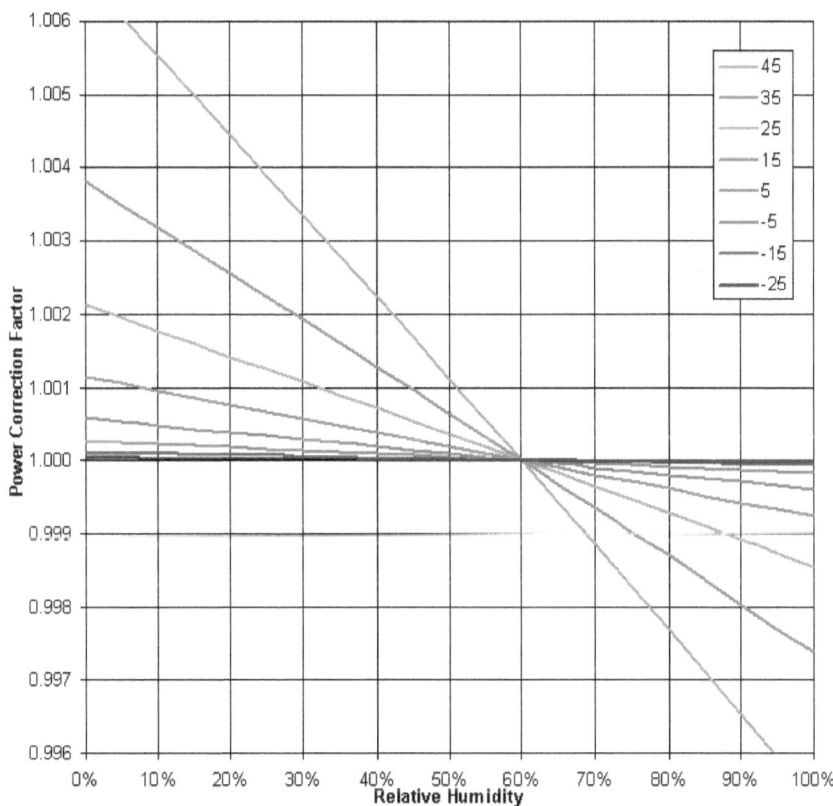

**Figure 31. Typical GT Power Correction for Relative Humidity**

The curves above are roughly linear. This is not always the case, even for the same engine. Moisture impacts combustion and also emissions so don't be surprised if the relative humidity curves for your engine are not linear. Curvature is not a problem, but the curves should definitely be continuous in slope. If they are not, consider a different engine. There is no shortage of manufacturers anxious to sell you their products. I would not consider an engine with performance curves that look like someone threw a plate of spaghetti against the wall and took a picture of it with their phone. I have a portfolio of such curves, which deserve a place in the Hall of Shame.

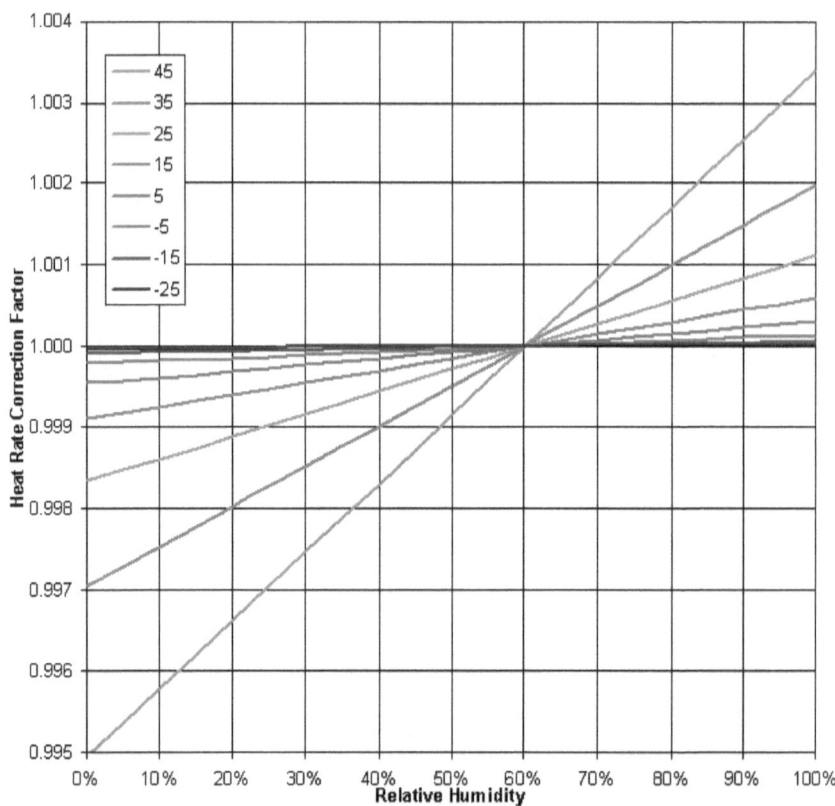

**Figure 32. Typical GT Heat Rate Correction for Relative Humidity**

Note that the slope of the power and heat rate corrections may be the same or opposite, depending on the engine, fuel, and tuning to meet emission regulations. Just because the corrections (except for temperature and barometric pressure) supplied by one manufacturer don't look like the ones you already have or are supplied by a different manufacturer, doesn't mean either one is wrong. Gas turbines are sufficiently complicated and expensive to build so that anyone manufacturing these must have considerable experience and expertise, which is not to say that all manufacturers are equal in every respect. This distinguishes gas turbines from simpler devices, which might be assembled from parts by someone who doesn't know what they're doing. Anyone who assembles a gas turbine and doesn't know what they're doing and is foolish enough to fire it up, will not likely survive to learn from their mistake.

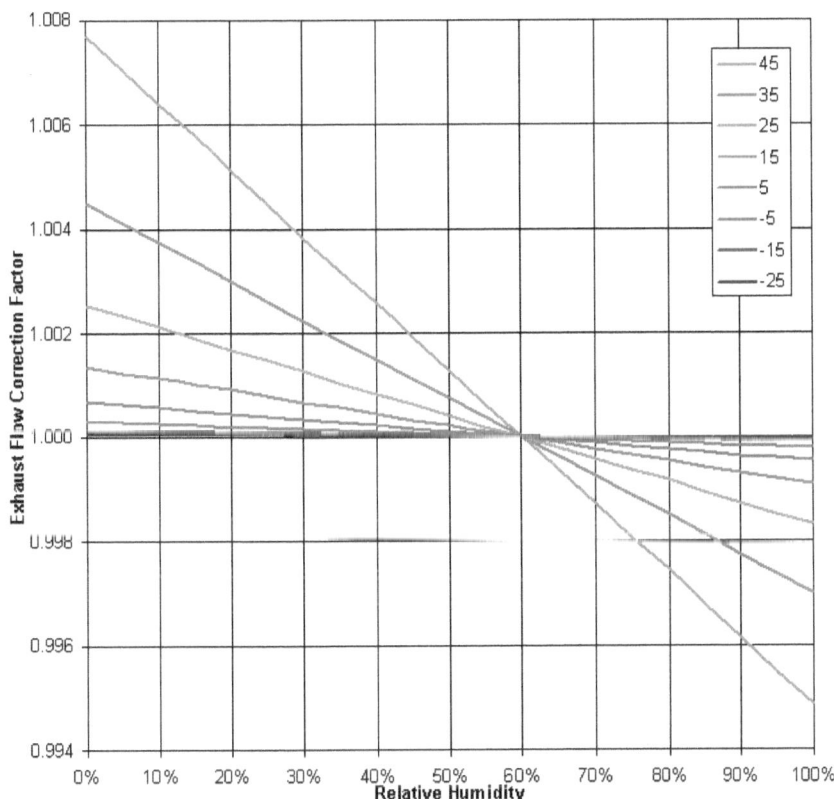

**Figure 33. Typical GT Exhaust Flow Correction for Relative Humidity**

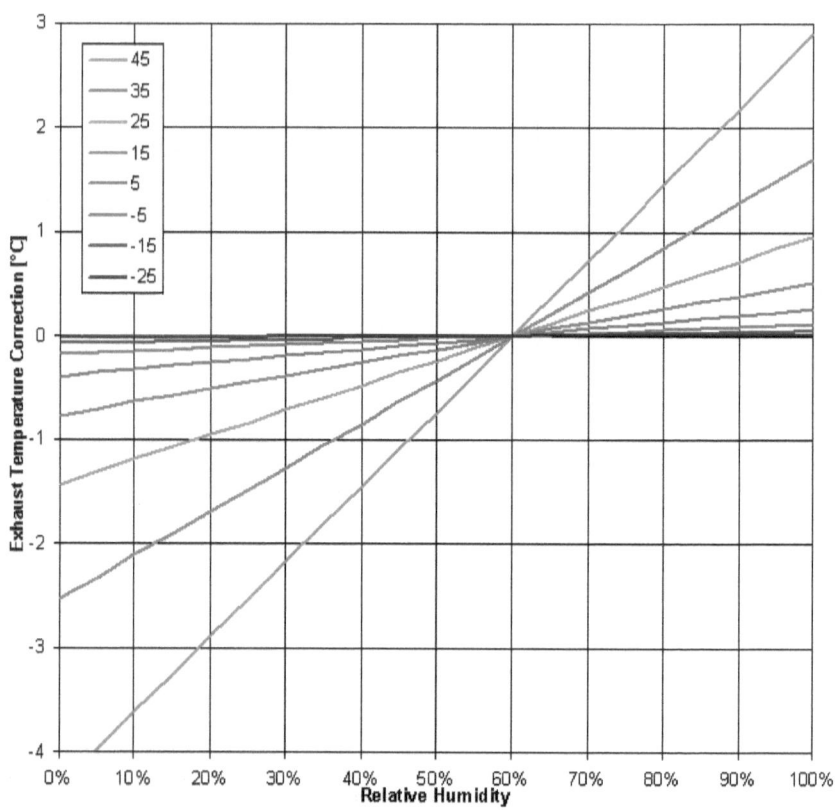

**Figure 34. Typical GT Exhaust Temperature Correction for RH**

### Absolute Humidity Corrections

While it requires a fan of curves to correct for relative humidity, these should all collapse to a single curve for absolute humidity. Relative humidity is the fraction (typically expressed as a percent) of the water bearing capacity of the air, which changes with temperature. Absolute humidity is the mass fraction of water vapor, or the mass of water vapor divided by the mass of water vapor plus the mass of dry air in some reference volume. It is the water *content* (i.e., absolute humidity), not the water *bearing capacity* of the air that impacts gas turbine performance, so this simplification should not be surprising. See Appendix A for more on relative and absolute humidity.

28

The figure below is the same information as shown on the previous four. All of these curves are easily normalized so that they go through any desired reference point.

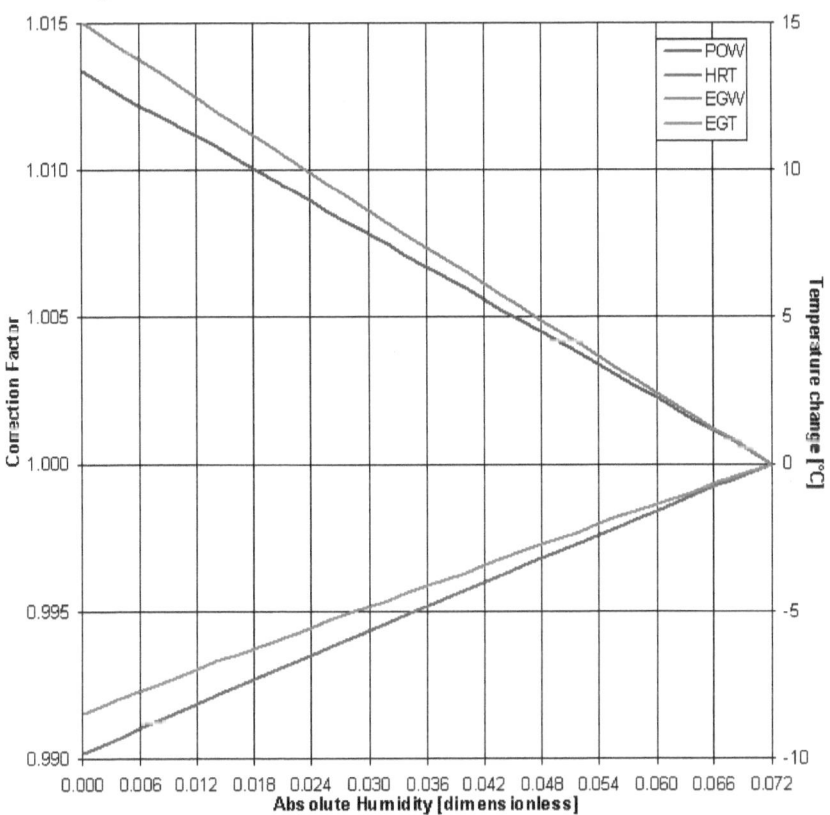

**Figure 35. Absolute Humidity Corrections**

Corrections for Fuel Composition

For sure, fuel composition does impact the performance of a gas turbine. The question is, "How much?" When you consider that these engines will run well enough on natural gas and also diesel—two vastly different fuels—how much difference could there be between 93.25% methane and 93.64%? Not only are these corrections relatively small compared to compressor inlet temperature and barometric pressure, they are purely empirical, that is, based on a scattering of data. Finding any fuel composition-related trend among the many other changing variables can be difficult. This is why only bi-linear corrections are used. As illustrated in Chapter 1, these typically appear as a stack of three straight lines.

29

The curves and data can be found in the online archive in folder examples in spreadsheet GT_curves.xls. Graphs and tables are on the tab, Sheet1. The source and regression are on tab, Sheet3. Data are most often provided in this format: four corrections (POW, HRT, EGW, EGT) at 3 different values of LHV at three different values of C/H. The three different values of LHV are not the same for each of the values of C/H, so this doesn't yield a 2D table from which you could calculate any value. The remedy is to perform a regression using the Excel function LINEST, which is illustrated on tab, Sheet3.

| correction data as supplied by manufacturer | | | | | |
|------|--------|---------|---------|---------|-------|
| C/H | LHV | power | ht.rt. | EGW | EGT |
| 4 | 23,891 | 0.99938 | 1.0002 | 0.99884 | 0.04 |
| 4 | 22,806 | 1.00122 | 0.99966 | 0.99994 | 0.16 |
| 4 | 21,794 | 1.00309 | 0.99914 | 1.00108 | 0.29 |
| 3.89 | 23,709 | 0.99827 | 1.0005 | 0.99897 | -0.11 |
| 3.89 | 22,704 | 1 | 1 | 1 | 0 |
| 3.89 | 21,739 | 1.0018 | 0.99949 | 1.00109 | 0.12 |
| 3.8 | 23,567 | 0.99722 | 1.0008 | 0.99906 | -0.24 |
| 3.8 | 22,610 | 0.99888 | 1.00031 | 1.00005 | -0.14 |
| 3.8 | 21,689 | 1.0006 | 0.99981 | 1.00109 | -0.03 |

**Figure 36. LHV and C/H Correction Data as Supplied**

| LINEST | | | correction |
|--------------|-----------|----------|---------------------|
| -1.78513E-06 | 0.013484 | 0.988074 | power |
| 5.14216E-07 | -0.00373 | 1.002832 | heat rate |
| -1.07416E-06 | 0.00055 | 1.022278 | exhaust flow |
| -0.000116125 | 1.619242 | -3.6609 | exhaust temperature |

**Figure 37. Bi-Linear Regression using LINEST**

| curve fit of power correction | | | | curve fit of heat rate correction | | | |
|--------|--------|--------|--------|--------|--------|--------|--------|
| | 3.80 | 3.89 | 4.00 | | 3.80 | 3.89 | 4.00 |
| 21,500 | 1.0009 | 1.0021 | 1.0036 | 21,500 | 0.9997 | 0.9994 | 0.9990 |
| 22,000 | 1.0000 | 1.0013 | 1.0027 | 22,000 | 1.0000 | 0.9996 | 0.9992 |
| 22,500 | 0.9991 | 1.0004 | 1.0018 | 22,500 | 1.0002 | 0.9999 | 0.9995 |
| 23,000 | 0.9983 | 0.9995 | 1.0010 | 23,000 | 1.0005 | 1.0002 | 0.9997 |
| 23,500 | 0.9974 | 0.9986 | 1.0001 | 23,500 | 1.0007 | 1.0004 | 1.0000 |
| 24,000 | 0.9965 | 0.9977 | 0.9992 | 24,000 | 1.0010 | 1.0007 | 1.0003 |

**Figure 38. Bi-Linear Tables for Power and Heat Rate**

| curve fit of ex. flow correction | | | | curve fit of ex. temp. correction | | |
|---|---|---|---|---|---|---|
|  | 3.80 | 3.89 | 4.00 |  | 3.80 | 3.89 | 4.00 |
| 21,500 | 1.0013 | 1.0013 | 1.0014 | 21,500 | 0.00 | 0.14 | 0.32 |
| 22,000 | 1.0007 | 1.0008 | 1.0008 | 22,000 | -0.06 | 0.08 | 0.26 |
| 22,500 | 1.0002 | 1.0002 | 1.0003 | 22,500 | -0.12 | 0.03 | 0.20 |
| 23,000 | 0.9997 | 0.9997 | 0.9998 | 23,000 | -0.18 | -0.03 | 0.15 |
| 23,500 | 0.9991 | 0.9992 | 0.9992 | 23,500 | -0.24 | -0.09 | 0.09 |
| 24,000 | 0.9986 | 0.9986 | 0.9987 | 24,000 | -0.29 | -0.15 | 0.03 |

**Figure 39. Bi-Linear Tables for Exhaust Flow and Temperature**

The curves themselves are similar to the ones for whole plant corrections shown in Chapter 1.

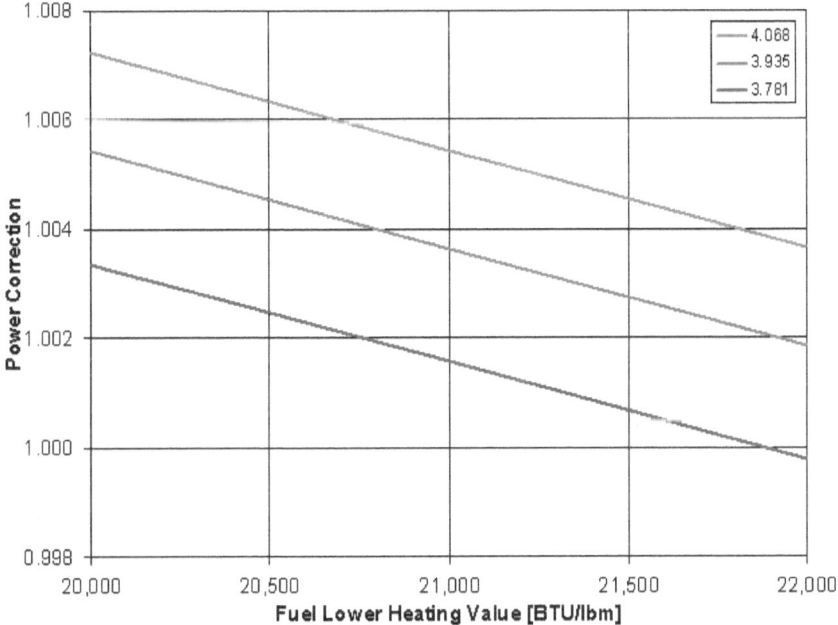

**Figure 40. GT Power Correction for Fuel Composition**

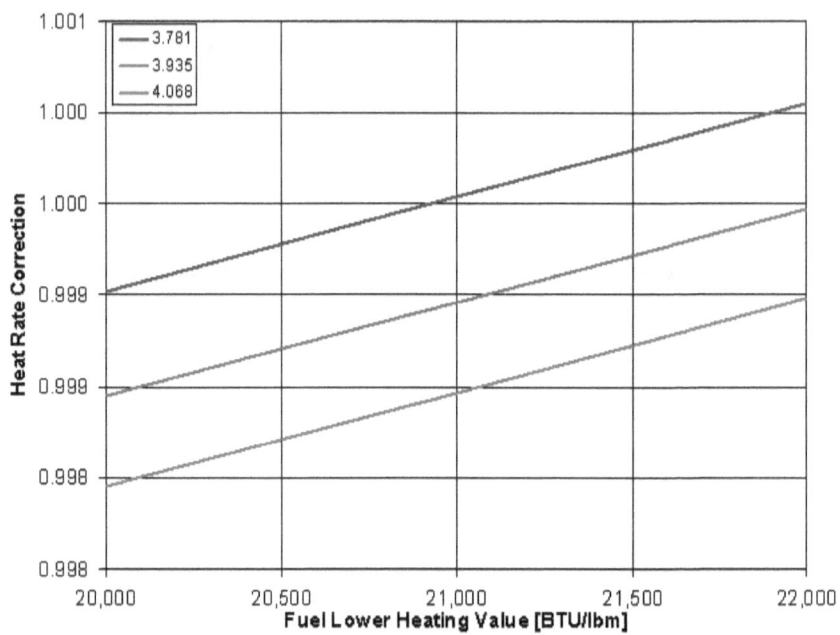

**Figure 41. GT Heat Rate Correction for Fuel Composition**

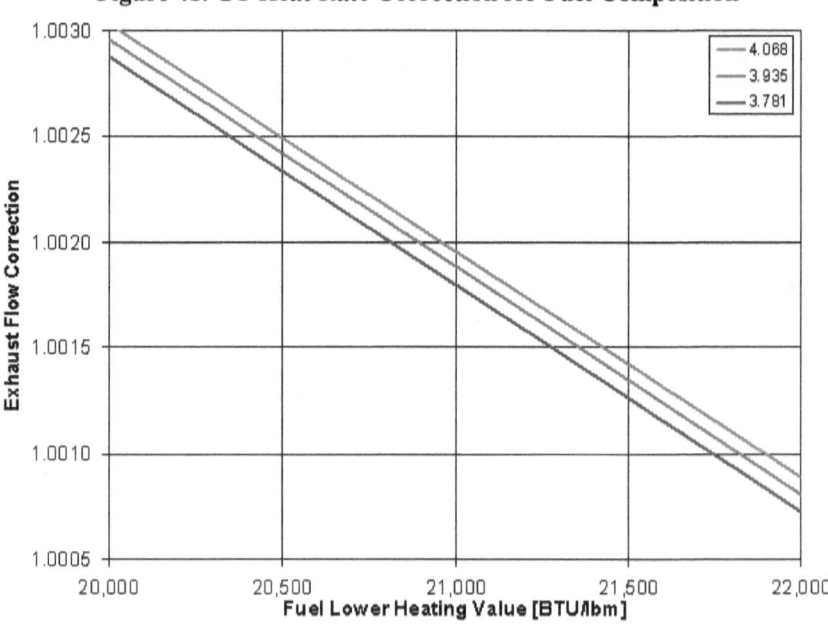

**Figure 42. GT Exhaust Flow Correction for Fuel Composition**

32

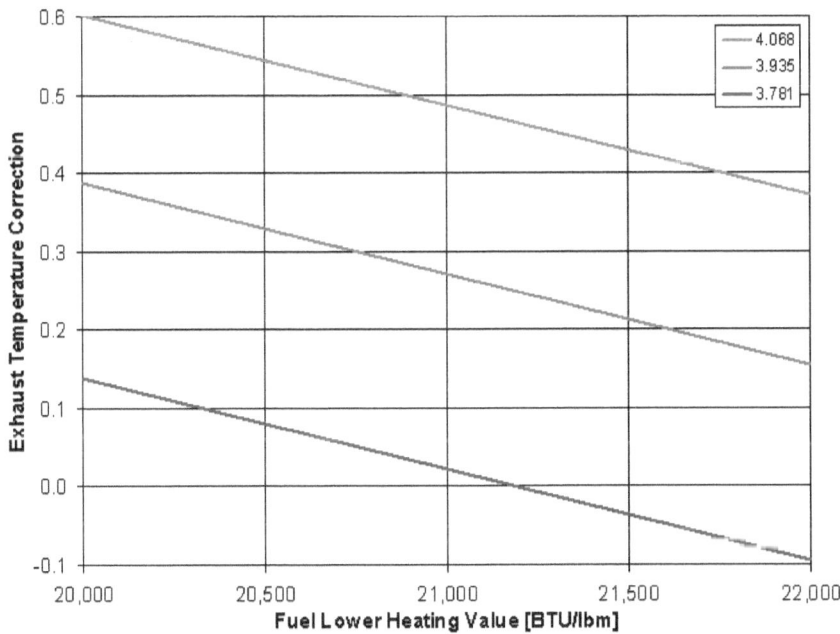

**Figure 43. GT Exhaust Temperature Correction for Fuel Composition**

<u>Adjustments for Aging</u>

Aging is generally quantified in terms of *equivalent operating hours*, or EOH. Various changes in operation (startup, shutdown, cleaning, etc.) are assigned some value, which differs with manufacturer. Corrections are often provided for power, heat rate, exhaust flow, and exhaust temperature. Typical curves are shown in the figure on the next page. These relationships can often be approximated by the power option trend line in Excel, which is the following equation:

$$y = 1 + at^b \qquad (2.1)$$

The exponent b is typically between 1/3 and 2/3.

33

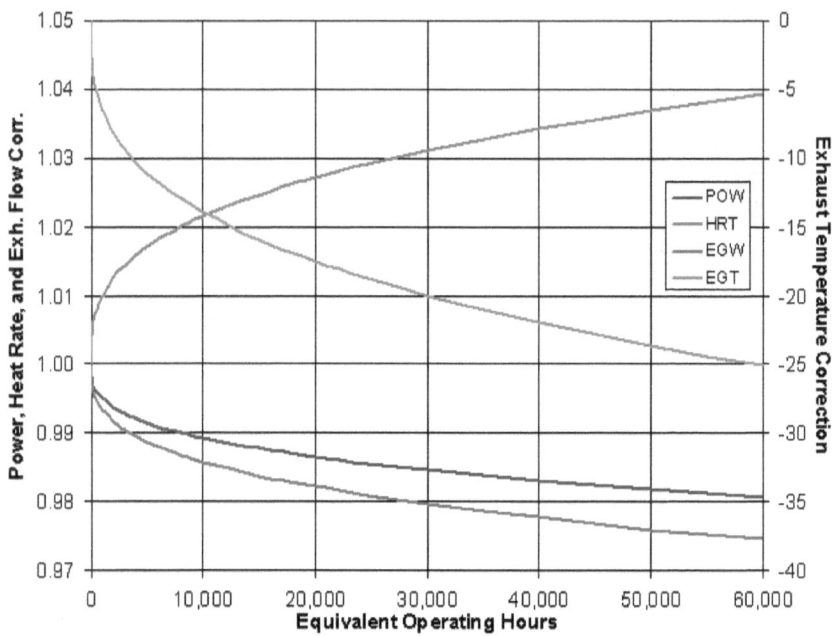

**Figure 44. GT Corrections for Aging**

<u>Applying the Corrections</u>

We will now consider how these corrections are applied, using the test example of the previous chapter. Here we will use SI units. As I stated before, the practicing scientist/engineer must be proficient with any and all systems of units. As these are *performance* curves, not *correction* curves, the inverse must be applied to the as-tested values. While it might not be true of all heat balance codes, GateCycle™ expects gas turbine performance to be entered as illustrated by the figures in this chapter (as *performance*, not *correction*). The curves and tables can be found in the online archive in the examples folder in spreadsheet GT_curves.xls. The curves are on tab, Sheet1, and the test calculations are on tab, Sheet2. Several functions are used for interpolation and curve-fitting. If you need these, you can get a free tool from my web site listed in the Forward. Look for the math Excel AddIn about halfway down on the page.

https://dudleybenton.altervista.org/software/index.html

Data, calculations, and results for a typical gas turbine test are shown in the tables on the next two pages.

34

| TYPICAL GAS TURBINE TEST CALCULATIONS | | | | | |
|---|---|---|---|---|---|
| TEST CONDITIONS | units | Test1 | Test2 | Test3 | Test4 |
| Barometric Pressure | kPa | 100.7 | 100.8 | 101.2 | 101.2 |
| Ambient Dry-Bulb Temperature | °C | 16.30 | 14.28 | 16.38 | 16.18 |
| Relative Humidity | - | 49% | 54% | 47% | 50% |
| Fuel Gas Heating Value | kJ/kg | 48,562 | 48,514 | 48,545 | 48,519 |
| Fuel Gas H/C Ratio | - | 3.805 | 3.803 | 3.805 | 3.805 |
| GT1 Equivaent Hours of Operation | EOH | 2728 | 2729 | 2730 | 2731 |
| GT2 Equivaent Hours of Operation | EOH | 2621 | 2622 | 2623 | 2624 |
| Power Factor | - | 0.891 | 0.892 | 0.893 | 0.891 |
| GT1 output | kWe | 85,818 | 86,719 | 86,133 | 86,164 |
| GT2 output | kWe | 85,521 | 86,569 | 85,984 | 86,015 |
| Fuel Flow to GT1 | kg/s | 5.314 | 5.372 | 5.344 | 5.348 |
| Fuel Flow to GT2 | kg/s | 5.332 | 5.391 | 5.363 | 5.367 |
| Steam Flow to GT1 | kg/s | 7.195 | 7.274 | 7.236 | 7.242 |
| Steam Flow to GT2 | kg/s | 7.680 | 7.775 | 7.130 | 7.150 |
| CALCULATIONS | units | Test1 | Test2 | Test3 | Test4 |
| GT1 Heat Input | kWt | 258,059 | 260,630 | 259,431 | 259,499 |
| GT2 Heat Input | kWt | 258,952 | 261,532 | 260,330 | 260,397 |
| GT1 Heat Rate | kJ/kWh | 10,825 | 10,820 | 10,843 | 10,842 |
| GT2 Heat Rate | kJ/kWh | 10,901 | 10,876 | 10,900 | 10,898 |
| ADDITIVE POWER CORRECTION | units | Test1 | Test2 | Test3 | Test4 |
| GT1 Power Factor | kW | -56 | -58 | -59 | -56 |
| GT2 Power Factor | kW | -55 | -58 | -58 | -56 |
| MULTIPLICATIVE POWER CORRECTIONS | units | Test1 | Test2 | Test3 | Test4 |
| Ambient Dry-Bulb Temperature | - | 1.0084 | 0.9957 | 1.0089 | 1.0077 |
| Barometric Pressure | - | 1.0068 | 1.0058 | 1.0012 | 1.0015 |
| Ambient Relative Humidity | - | 0.9976 | 0.9990 | 0.9973 | 0.9978 |
| GT1 Steam Power Augmentation | - | 1.0014 | 1.0014 | 1.0014 | 1.0014 |
| GT2 Steam Power Augmentation | - | 0.9928 | 0.9926 | 1.0038 | 1.0035 |
| Fuel Gas LHV and H/C | - | 0.9979 | 0.9979 | 0.9979 | 0.9979 |
| GT1 Equivaent Hours of Operation | - | 1.0070 | 1.0070 | 1.0070 | 1.0070 |
| GT2 Equivaent Hours of Operation | - | 1.0069 | 1.0069 | 1.0069 | 1.0069 |
| Total Multiplicative Power Correction for GT1 | - | 1.0191 | 1.0067 | 1.0137 | 1.0133 |
| Total Multiplicative Power Correction for GT2 | - | 1.0104 | 0.9979 | 1.0160 | 1.0154 |

Figure 45. Typical Gas Turbine Test Calculations

| MULTIPLICATIVE HEAT RATE CORRECTIONS | units | Test1 | Test2 | Test3 | Test4 |
|---|---|---|---|---|---|
| Ambient Dry-Bulb Temperature | - | 0.9982 | 1.0009 | 0.9981 | 0.9984 |
| Barometric Pressure | - | 0.9999 | 0.9999 | 1.0000 | 1.0000 |
| Ambient Relative Humidity | - | 1.0007 | 1.0003 | 1.0008 | 1.0006 |
| GT1 Steam Power Augmentation | - | 1.0007 | 1.0007 | 1.0007 | 1.0007 |
| GT2 Steam Power Augmentation | - | 0.9962 | 0.9961 | 1.0020 | 1.0018 |
| Fuel Gas LHV and H/C | - | 1.0006 | 1.0006 | 1.0006 | 1.0006 |
| GT1 Equivaent Hours of Operation | - | 1.0139 | 1.0139 | 1.0139 | 1.0139 |
| GT2 Equivaent Hours of Operation | - | 1.0137 | 1.0137 | 1.0137 | 1.0137 |
| Total Multiplicative Heat Rate Correction for GT1 | | 1.0140 | 1.0164 | 1.0141 | 1.0142 |
| Total Multiplicative Heat Rate Correction for GT2 | | 1.0093 | 1.0115 | 1.0152 | 1.0152 |
| CORRECTED TEST VALUES | units | Test1 | Test2 | Test3 | Test4 |
| GT1 Power | kW | 87,404 | 87,243 | 87,250 | 87,250 |
| GT2 Power | kW | 86,352 | 86,326 | 87,303 | 87,280 |
| GT1 Heat Rate | kJ/kWh | 10,977 | 10,997 | 10,996 | 10,996 |
| GT2 Heat Rate | kJ/kWh | 11,002 | 11,001 | 11,065 | 11,064 |
| CORRECTED TEST MARGINS | units | Test1 | Test2 | Test3 | Test4 |
| GT1 Power | - | 0.28% | 0.10% | 0.10% | 0.10% |
| GT2 Power | - | -0.93% | -0.96% | 0.16% | 0.14% |
| GT1 Heat Rate | - | -0.23% | -0.05% | -0.05% | -0.05% |
| GT2 Heat Rate | - | 0.00% | -0.01% | 0.58% | 0.56% |
| TEST RESULTS | units | Test1 | Test2 | Test3 | Test4 |
| GT1 Power | - | PASS | PASS | PASS | PASS |
| GT2 Power | - | FAIL | FAIL | PASS | PASS |
| GT1 Heat Rate | - | PASS | PASS | PASS | PASS |
| GT2 Heat Rate | - | PASS | PASS | FAIL | FAIL |

**Figure 46. Typical Gas Turbine Test Results**

## Chapter 3. Expected Performance

We can use these curves along with weather data to predict the performance of a plant for the purposes of dispatch, that is, scheduling available capacity to meet demand. I explain elsewhere how to acquire and utilize meteorological data.[6] We will begin with a typical year of data, which you can find in the online archive in folder examples and spreadsheet simulation1.xls.

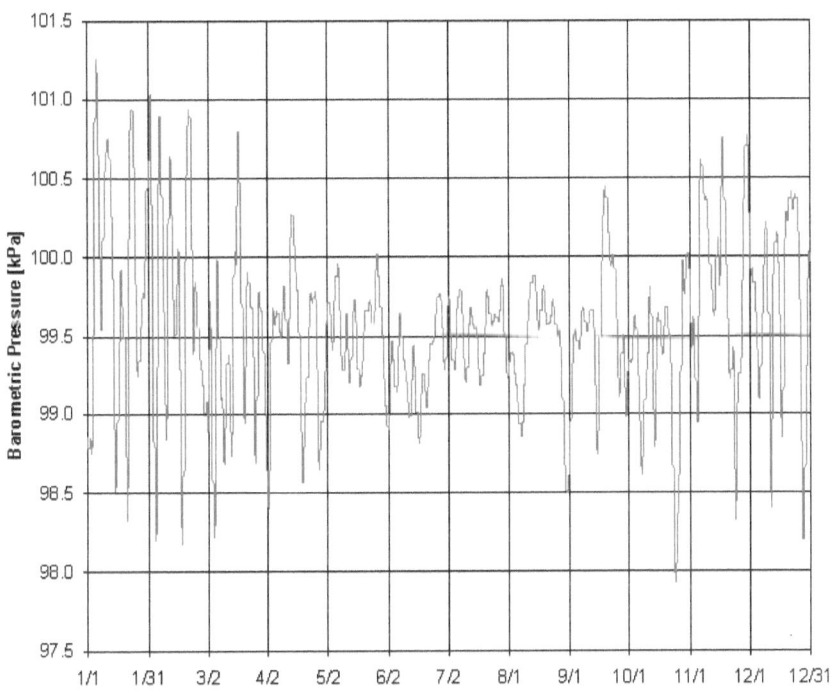

**Figure 47. Hourly Values of Barometric Pressure**

---

[6] *Computer Simulation of Power Systems:* Programming Strategies and Practical Examples, ISBN-9781696218184, Amazon 2019. See Appendices B and C. The software and examples accompanying this book are available free online at the same location listed in the Forward. There is also a list there of when the Ebook is free.

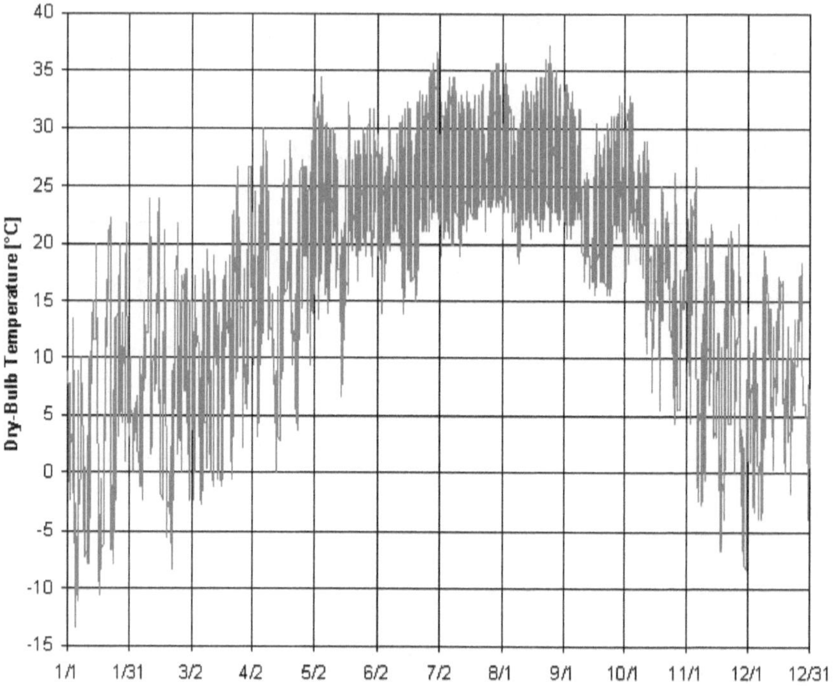

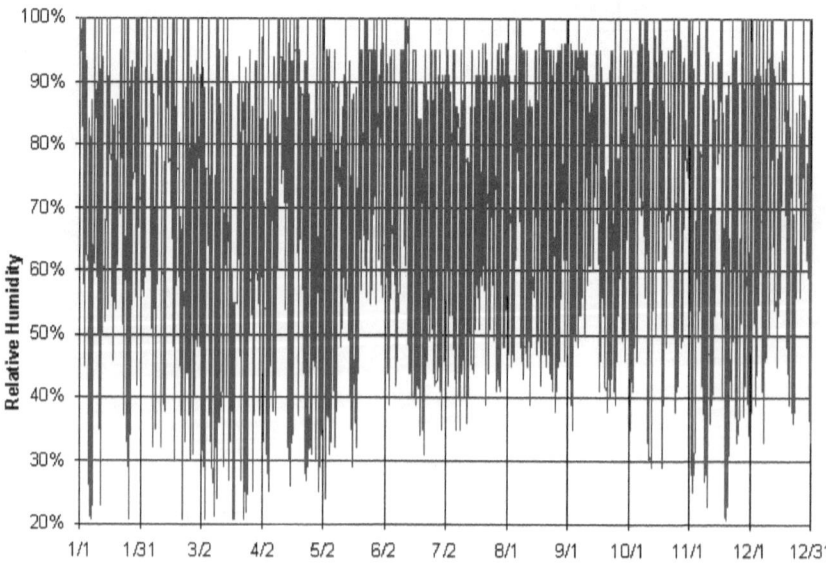

**Figure 48. Hourly Values of Dry-Bulb Temperature**

**Figure 49. Hourly Values of Relative Humidity**

In the next figures we see that the fuel heating value does vary significantly throughout the year.

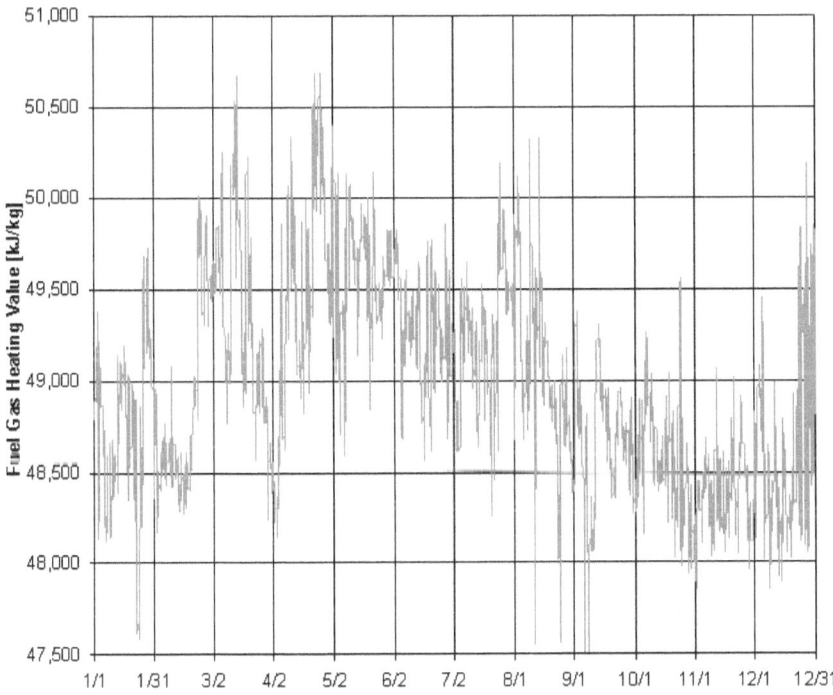

**Figure 50. Hourly Values of Heating Value from Online GG**

While the heating value of this particular fuel does vary, we see in this next figure that the carbon-to-hydrogen ratio does not. This is because the components varying throughout the year are the non-hydrocarbons, mostly carbon dioxide and nitrogen.

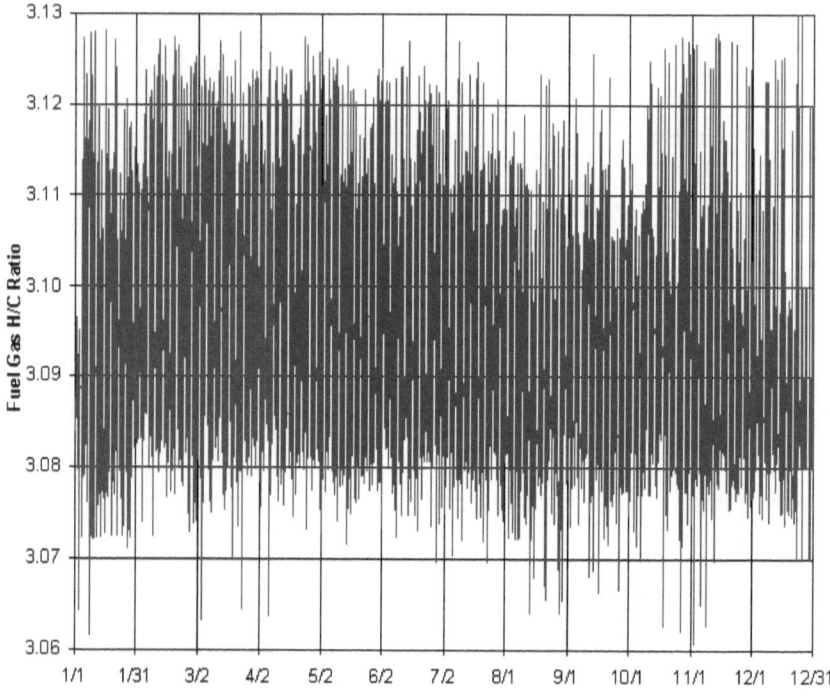

**Figure 51. Hourly Values of C/H Ratio from Online GC**

The magnitude of these last two variations convey a sense of the relative importance of fuel composition measurements and corrections.

## Performing the Simulation

We begin with the GT corrections from spreadsheet GT_curves.xls, perform regressions as needed[7], convert these to Excel macros, copy them into

---

[7] Regression tools are available free at the link listed in the Forward. CurveFit can perform single or multivariate regressions and will create functions that can be pasted directly into Excel so you need not worry about copying the coefficients correctly. Also notice that there's an option to force the regression to match exactly at one of the data points. Use this to fit the reference conditions.

simulation1.xls, and finally calculate the expected power output. The required macros are listed below:

```
Option Explicit
Function POWCIT(CIT As Double) As Double
  POWCIT = (-4.56084030179E-06 * CIT - 0.00607220994284)
    * CIT + 1.09210933821
End Function
Function HRTCIT(CIT As Double) As Double
  HRTCIT = (1.06189437587E-05 * CIT + 0.000964667835046)
    * CIT + 0.983140720129
End Function
Function POWBAR(baro As Double) As Double
  POWBAR = 0.0101156222062 * baro - 0.0252479716308
End Function
Function HRTBAR(baro As Double) As Double
  HRTBAR = -0.000189586949039 * baro + 1.01021519319
End Function
Function POWRH(RH As Double, CIT As Double) As Double
  POWRH = 0.998007050124 + 8.00335771673E-06 * CIT ^ 2 -
    1.29064666545E-05 * RH * CIT ^ 2
End Function
Function HRTRH(RH As Double, CIT As Double) As Double
  HRTRH = 1.00030076836 - 1.70056134321E-06 * CIT ^ 2 +
    2.7713275993E-06 * RH * CIT ^ 2
End Function
Function POWEOH(EOH As Double) As Double
  POWEOH = 1 / (1 + (EOH ^ 0.333333) / 2000)
End Function
Function POWLHV(LHV As Double, CH As Double) As Double
  POWLHV = -1.7851332218922E-06 * LHV +
    1.34837131765734E-02 * CH + 0.988073554330391
End Function
Function HRTLHV(LHV As Double, CH As Double) As Double
  HRTLHV = 5.14216033118933E-07 * LHV -
    3.7280622408577E-03 * CH + 1.00283237067442
End Function
Function HRTEOH(EOH As Double) As Double
  HRTEOH = 1 + (EOH ^ 0.333333) / 1000
End Function
Function POWEXP(baro As Double, CIT As Double, RH As
    Double, CH As Double, LHV As Double, EOH As Double)
    As Double
  POWEXP = 83500 * POWCIT(CIT) * POWBAR(baro) *
    POWRH(RH, CIT) / POWRH(0.6, CIT) _
    * POWLHV(LHV / 2.326, CH) / POWLHV(21500, 3.935) *
    POWEOH(EOH)
End Function
```

41

```
Function HRTEXP(baro As Double, CIT As Double, RH As
    Double, CH As Double, LHV As Double, EOH As Double)
    As Double
  HRTEXP = 10480 * HRTCIT(CIT) * HRTBAR(baro) *
    HRTRH(RH, CIT) / HRTRH(0.6, CIT) _
    * HRTLHV(LHV / 2.326, CH) / HRTLHV(21500, 3.935) *
    HRTEOH(EOH) * 3600 / 3412.141633
End Function
```

The expected power is the rated value for the engine (in this case 83,500 kW) times each of the corrections. The temperature, barometric pressure, and aging corrections have been normalized. The relative humidity and fuel composition corrections have not been normalized, so we divide by the reference values as indicated below:

```
*POWRH(RH,CIT)/POWRH(0.6,CIT)
*POWLHV(LHV/2.326,CH)/POWLHV(21500,3.935)
```

The fuel composition corrections here include a unit conversion from kJ/kg to BTU/lbm. This illustrates how the practicing scientist/engineer must be familiar with units, conversions, and typical values of the quantities in question. I have received data on many occasions without units or with the wrong units and so knowing what reasonable values are is essential. Huge mistakes can result from ignorance of these matters, like JPL missing Mars in 1999 and wasting $125M.

The expected heat rate is calculated in similar fashion from the rated value (in this case 10480 time the unit conversion 3600/3412) times the corrections, with the same two (relative humidity and fuel composition) normalized.

The following table shows the input data (atmospheric conditions, fuel composition, and engine age) and the output calculations (power, heat rate, and fuel flow). SI units are shown at the top of each column.

Hourly Simulation of Typical 2x0 Simple Cycle Power Plant Using Historical Weather Data

| m/d/yyyy h:mm | baro kPa | Tdb °C | RH % | H/C - | LHV kJ/kg | EOH1 hours | EOH2 hours | POW1 kW | POW2 kW | HRT1 kJ/kWh | HRT2 kJ/kWh | fuel1 kg/hr | fuel2 kg/hr |
|---|---|---|---|---|---|---|---|---|---|---|---|---|---|
| 1/1/2019 0:00 | 98.87 | 8.89 | 87% | 3.08 | 48,840 | 8,409 | 9,117 | 82,722 | 82,699 | 11,236 | 11,243 | 19,032 | 19,037 |
| 1/1/2019 1:00 | 98.85 | 8.89 | 87% | 3.09 | 48,842 | 8,410 | 9,118 | 82,707 | 82,684 | 11,236 | 11,243 | 19,027 | 19,032 |
| 1/1/2019 2:00 | 98.83 | 8.89 | 87% | 3.09 | 48,831 | 8,412 | 9,120 | 82,692 | 82,669 | 11,236 | 11,243 | 19,028 | 19,033 |
| 1/1/2019 3:00 | 98.82 | 8.89 | 87% | 3.09 | 48,833 | 8,413 | 9,121 | 82,683 | 82,660 | 11,236 | 11,243 | 19,025 | 19,030 |
| 1/1/2019 4:00 | 98.81 | 8.33 | 93% | 3.09 | 48,832 | 8,414 | 9,122 | 82,946 | 82,924 | 11,229 | 11,235 | 19,074 | 19,079 |
| 1/1/2019 5:00 | 98.80 | 7.78 | 100% | 3.09 | 48,815 | 8,415 | 9,123 | 83,205 | 83,183 | 11,222 | 11,228 | 19,128 | 19,134 |
| 1/1/2019 6:00 | 98.79 | 7.78 | 93% | 3.08 | 48,798 | 8,416 | 9,124 | 83,201 | 83,178 | 11,222 | 11,228 | 19,134 | 19,139 |
| 1/1/2019 7:00 | 98.79 | 7.78 | 93% | 3.08 | 48,807 | 8,417 | 9,125 | 83,200 | 83,177 | 11,222 | 11,228 | 19,130 | 19,135 |
| 1/1/2019 8:00 | 98.79 | 7.78 | 100% | 3.08 | 48,846 | 8,418 | 9,126 | 83,194 | 83,171 | 11,222 | 11,229 | 19,114 | 19,119 |
| 1/1/2019 9:00 | 98.79 | 7.78 | 93% | 3.08 | 48,853 | 8,419 | 9,127 | 83,198 | 83,175 | 11,222 | 11,228 | 19,112 | 19,117 |
| 1/1/2019 10:00 | 98.79 | 7.78 | 100% | 3.08 | 48,887 | 8,420 | 9,128 | 83,192 | 83,169 | 11,223 | 11,229 | 19,098 | 19,103 |
| 1/1/2019 11:00 | 98.79 | 7.22 | 100% | 3.08 | 48,914 | 8,421 | 9,129 | 83,471 | 83,448 | 11,215 | 11,221 | 19,139 | 19,144 |
| 1/1/2019 12:00 | 98.79 | 6.67 | 100% | 3.09 | 48,890 | 8,422 | 9,130 | 83,745 | 83,722 | 11,208 | 11,214 | 19,199 | 19,204 |
| 1/1/2019 13:00 | 98.79 | 6.11 | 93% | 3.09 | 48,912 | 8,423 | 9,131 | 84,024 | 84,001 | 11,201 | 11,207 | 19,242 | 19,247 |
| 1/1/2019 14:00 | 98.80 | 5.56 | 100% | 3.09 | 48,925 | 8,424 | 9,132 | 84,300 | 84,276 | 11,194 | 11,200 | 19,288 | 19,293 |
| 1/1/2019 15:00 | 98.81 | 5.56 | 92% | 3.09 | 48,977 | 8,425 | 9,133 | 84,308 | 84,285 | 11,194 | 11,200 | 19,270 | 19,275 |
| 1/1/2019 16:00 | 98.82 | 5.56 | 92% | 3.09 | 49,003 | 8,426 | 9,134 | 84,316 | 84,293 | 11,194 | 11,200 | 19,261 | 19,266 |
| 1/1/2019 17:00 | 98.82 | 5.56 | 85% | 3.09 | 49,072 | 8,427 | 9,135 | 84,318 | 84,295 | 11,194 | 11,200 | 19,235 | 19,240 |
| 1/1/2019 18:00 | 98.83 | 5.00 | 85% | 3.10 | 49,115 | 8,428 | 9,136 | 84,600 | 84,577 | 11,187 | 11,194 | 19,270 | 19,276 |
| 1/1/2019 19:00 | 98.83 | 4.44 | 84% | 3.10 | 49,130 | 8,429 | 9,137 | 84,874 | 84,850 | 11,181 | 11,187 | 19,315 | 19,320 |
| 1/1/2019 20:00 | 98.84 | 4.44 | 84% | 3.09 | 49,156 | 8,430 | 9,138 | 84,880 | 84,857 | 11,181 | 11,187 | 19,306 | 19,312 |
| 1/1/2019 21:00 | 98.84 | 3.33 | 83% | 3.10 | 49,160 | 8,431 | 9,139 | 85,424 | 85,401 | 11,168 | 11,174 | 19,406 | 19,411 |
| 1/1/2019 22:00 | 98.84 | 1.67 | 100% | 3.10 | 49,154 | 8,432 | 9,140 | 86,233 | 86,209 | 11,148 | 11,154 | 19,558 | 19,563 |
| 1/1/2019 23:00 | 98.85 | 1.11 | 100% | 3.10 | 49,174 | 8,433 | 9,141 | 86,512 | 86,489 | 11,142 | 11,148 | 19,602 | 19,608 |

**Figure 52. Hourly Simulation of Typical 2x0 Simple Cycle Plant**

In this next figure we see that there is considerable change in capacity during the year (higher in the winter and lower in the summer). The capacity of GT1 and GT2 are quite similar in this case, which is not uncommon. The two lines are essentially on top of each other and not worth showing separately.

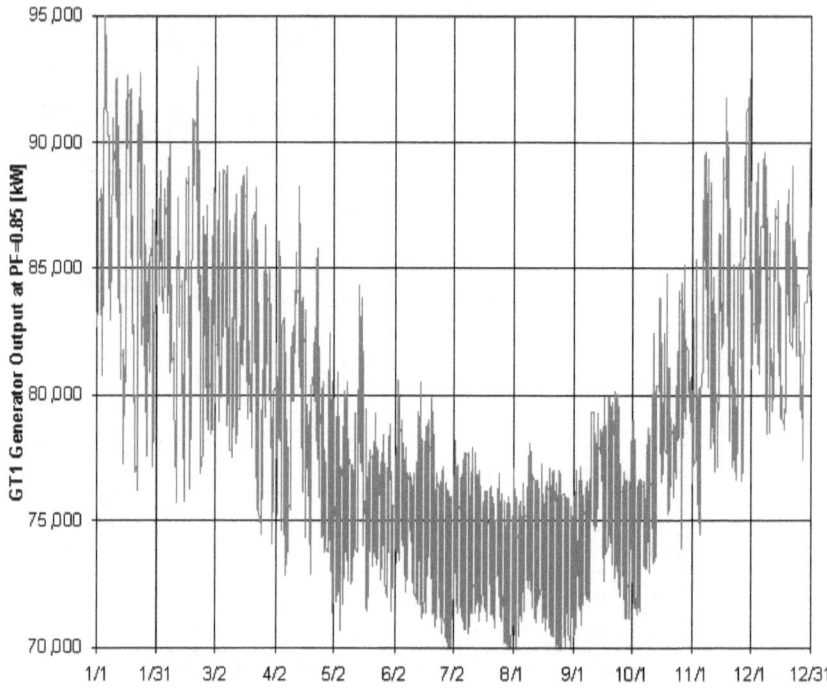

**Figure 53. Expected GT1 Capacity throughout the Year**

In this next figure we see that the heat rate is almost a mirror image of the capacity. This is as expected and is also borne out in the curves shown previously.

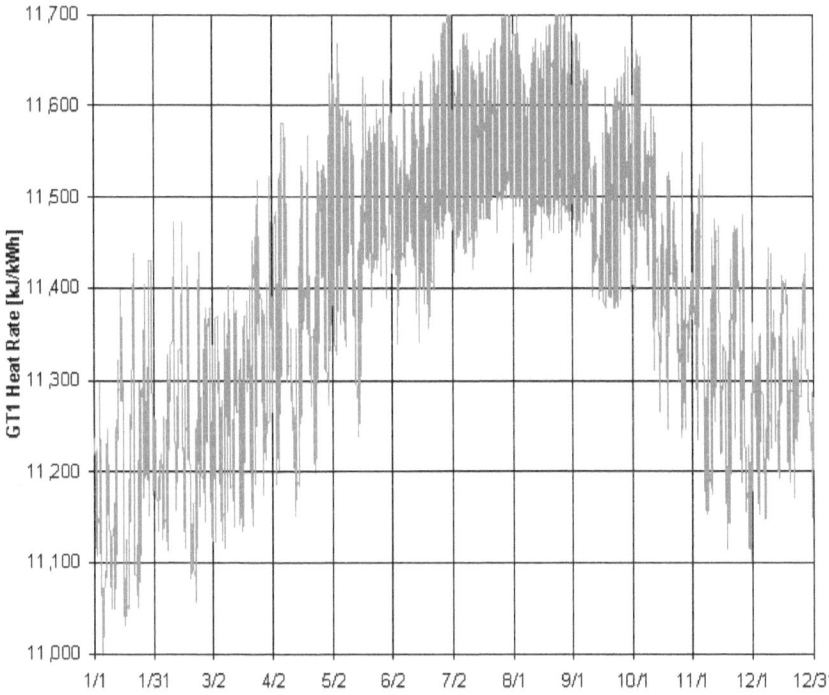

**Figure 54. Expected GT1 Heat Rate throughout the Year**

In this next figure we see that the fuel flow is the product of the power and heat rate divided by the heating value. The fuel flow is highest in the winder and lowest in the summer months.

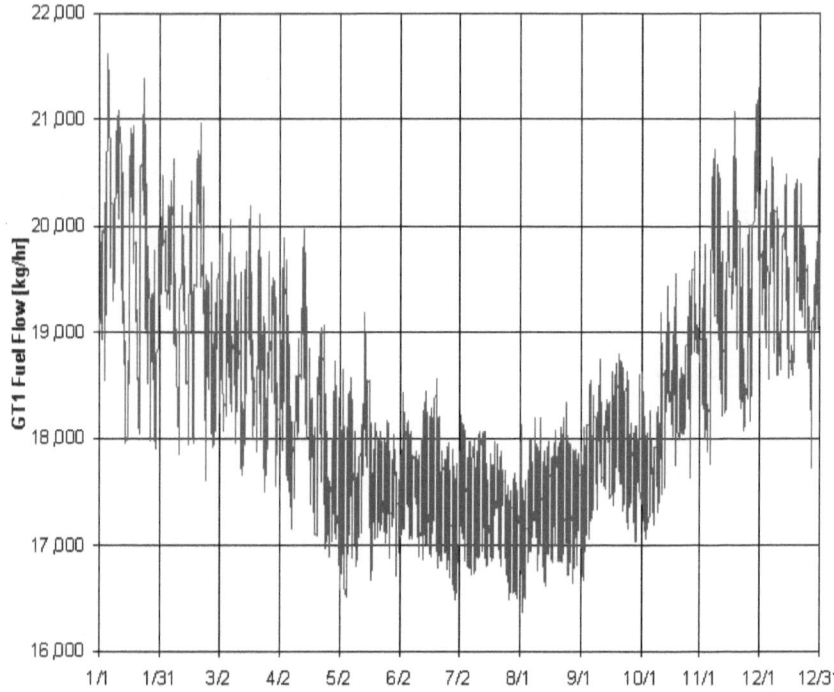

**Figure 55. Expected GT1 Fuel Flow throughout the Year**

The capacity (ability to produce power) is lowest in the summer and the heat rate (inverse of the efficiency at which power is produced) is highest in the summer, which means that supply is lowest in the summer and unit cost is highest. Of course, the summer and winter months are swapped above and below the equator.

## Combined Cycle Performance

We next consider the expected performance of a 2x1 CCPP. We will use the same weather data. The corrections were presented in Chapter 1. With a combined cycle plant we have the added versatility and capacity of duct firing. Power demand is highest in the winter for heating and the summer for cooling. Demand is lowest in the spring and fall. The capacity of our simple cycle was highest in the winter and lowest in the summer. In this simulation we will use duct firing to increase the summer output and better follow the demand. The

46

calculations and graphs are in spreadsheet simulation2.xls in the examples folder. The simulation spreadsheet looks like:

| Simulation of Typical 2x1 Combined Cycle Power Plant | | | | | | | | |
|---|---|---|---|---|---|---|---|---|
| m/d/yyyy | baro | Tdb | RH | Duct | Power | Heat In | Ht. Rt. | fuel |
| h:mm | kPa | °C | % | Firing | kWe | kWt | kJ/kWh | kg/hr |
| 1/1/2019 0:00 | 98.87 | 8.89 | 87% | FALSE | 245,006 | 521,230 | 7,659 | 38,373 |
| 1/1/2019 1:00 | 98.85 | 8.89 | 87% | FALSE | 244,958 | 521,124 | 7,659 | 38,365 |
| 1/1/2019 2:00 | 98.83 | 8.89 | 87% | FALSE | 244,910 | 521,019 | 7,659 | 38,357 |
| 1/1/2019 3:00 | 98.82 | 8.89 | 87% | FALSE | 244,887 | 520,966 | 7,659 | 38,353 |
| 1/1/2019 4:00 | 98.81 | 8.33 | 93% | FALSE | 245,185 | 521,650 | 7,659 | 38,404 |
| 1/1/2019 5:00 | 98.80 | 7.78 | 100% | FALSE | 245,448 | 522,243 | 7,660 | 38,447 |
| 1/1/2019 6:00 | 98.79 | 7.78 | 93% | FALSE | 245,735 | 522,989 | 7,662 | 38,502 |
| 1/1/2019 7:00 | 98.79 | 7.78 | 93% | FALSE | 245,735 | 522,989 | 7,662 | 38,502 |
| 1/1/2019 8:00 | 98.79 | 7.78 | 100% | FALSE | 245,424 | 522,190 | 7,660 | 38,443 |
| 1/1/2019 9:00 | 98.79 | 7.78 | 93% | FALSE | 245,735 | 522,989 | 7,662 | 38,502 |
| 1/1/2019 10:00 | 98.79 | 7.78 | 100% | FALSE | 245,424 | 522,190 | 7,660 | 38,443 |
| 1/1/2019 11:00 | 98.79 | 7.22 | 100% | FALSE | 246,043 | 523,687 | 7,662 | 38,554 |
| 1/1/2019 12:00 | 98.79 | 6.67 | 100% | FALSE | 246,647 | 525,150 | 7,665 | 38,661 |
| 1/1/2019 13:00 | 98.79 | 6.11 | 93% | FALSE | 247,533 | 527,337 | 7,669 | 38,822 |
| 1/1/2019 14:00 | 98.80 | 5.56 | 100% | FALSE | 247,882 | 528,135 | 7,670 | 38,881 |
| 1/1/2019 15:00 | 98.81 | 5.56 | 92% | FALSE | 248,206 | 528,960 | 7,672 | 38,942 |
| 1/1/2019 16:00 | 98.82 | 5.56 | 92% | FALSE | 248,230 | 529,014 | 7,672 | 38,946 |
| 1/1/2019 17:00 | 98.82 | 5.56 | 85% | FALSE | 248,493 | 529,691 | 7,674 | 38,996 |
| 1/1/2019 18:00 | 98.83 | 5.00 | 85% | FALSE | 249,099 | 531,151 | 7,676 | 39,103 |
| 1/1/2019 19:00 | 98.83 | 4.44 | 84% | FALSE | 249,713 | 532,642 | 7,679 | 39,213 |
| 1/1/2019 20:00 | 98.84 | 4.44 | 84% | FALSE | 249,737 | 532,696 | 7,679 | 39,217 |
| 1/1/2019 21:00 | 98.84 | 3.33 | 83% | FALSE | 250,909 | 535,541 | 7,684 | 39,426 |
| 1/1/2010 22:00 | 90.04 | 1.67 | 100% | FALSE | 252,149 | 538,497 | 7,688 | 39,644 |
| 1/1/2019 23:00 | 98.85 | 1.11 | 100% | FALSE | 252,768 | 540,007 | 7,691 | 39,755 |
| 1/2/2019 0:00 | 98.85 | 0.00 | 100% | FALSE | 253,942 | 542,901 | 7,696 | 39,968 |
| 1/2/2019 1:00 | 98.85 | 0.00 | 90% | FALSE | 254,165 | 543,478 | 7,698 | 40,011 |
| 1/2/2019 2:00 | 98.85 | 0.00 | 100% | FALSE | 253,942 | 542,901 | 7,696 | 39,968 |
| 1/2/2019 3:00 | 98.84 | 0.00 | 100% | FALSE | 253,917 | 542,846 | 7,696 | 39,964 |
| 1/2/2019 4:00 | 98.84 | -0.56 | 100% | FALSE | 254,509 | 544,313 | 7,699 | 40,072 |
| 1/2/2019 5:00 | 98.84 | -1.11 | 100% | FALSE | 255,091 | 545,759 | 7,702 | 40,179 |
| 1/2/2019 6:00 | 98.83 | -1.67 | 100% | FALSE | 255,658 | 547,183 | 7,705 | 40,283 |
| 1/2/2019 7:00 | 98.82 | -2.22 | 89% | FALSE | 256,405 | 549,082 | 7,709 | 40,423 |
| 1/2/2019 8:00 | 98.82 | -1.67 | 100% | FALSE | 255,633 | 547,128 | 7,705 | 40,279 |
| 1/2/2019 9:00 | 98.81 | 0.00 | 100% | FALSE | 253,843 | 542,601 | 7,696 | 39,952 |
| 1/2/2019 10:00 | 98.80 | 2.78 | 83% | FALSE | 251,372 | 536,688 | 7,686 | 39,511 |
| 1/2/2019 11:00 | 98.79 | 3.89 | 76% | FALSE | 250,443 | 534,470 | 7,683 | 39,348 |
| 1/2/2019 12:00 | 98.79 | 5.56 | 70% | FALSE | 248,986 | 530,986 | 7,677 | 39,091 |

**Figure 56. Hourly Simulation of Typical 2x1 Combined Cycle Plant**

47

Without duct firing our simulation predicts the following capacity:

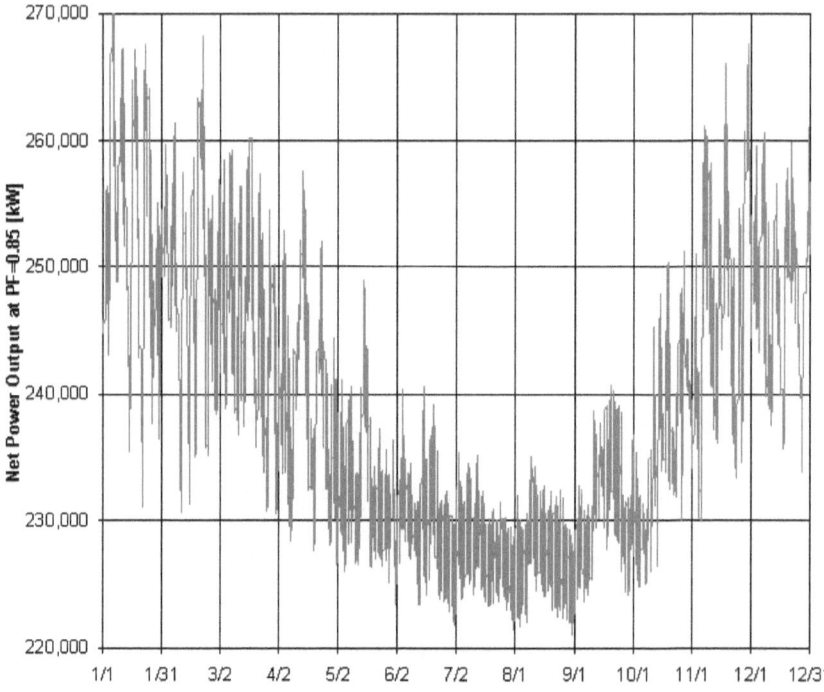

**Figure 57. 2x1 CCPP Capacity without Duct Firing**

A very simple test for when to activate duct firing is implemented in this simulation spreadsheet. If the ambient temperature is above some specified value, then duct firing is activated. The value is in cell J1. While there is some impact of fuel composition, as we saw in the last example, further illustration is unnecessary, so that information has been removed from simulation2.xls and a single average value for heating value is used. That value is in cell J2. You can change either of these parameters and the simulation will automatically update.

If we activate duct firing when the ambient temperature is above 20°C, we obtain the following predicted capacity:

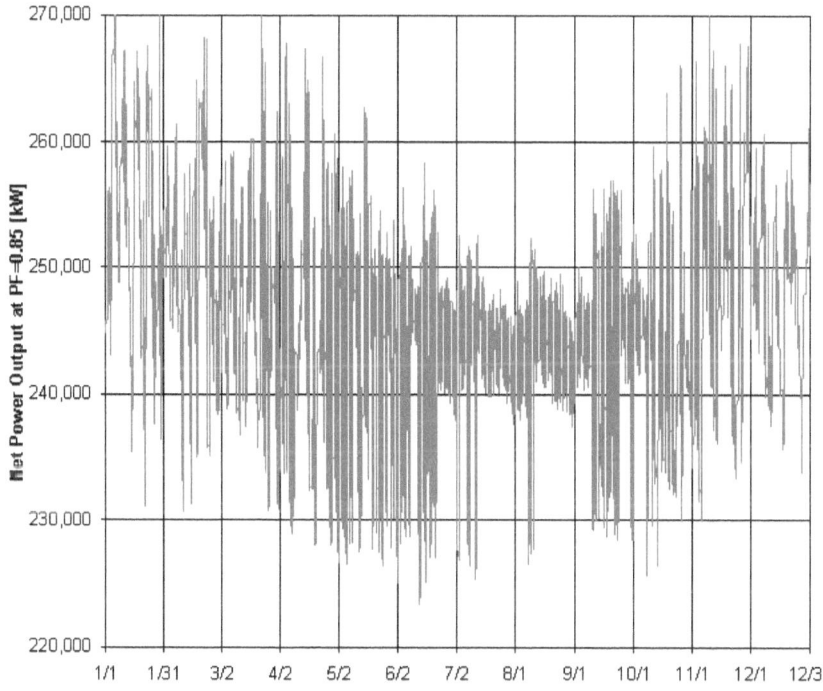

**Figure 58. 2x1 CCPP Capacity with Seasonal Duct Firing**

The heat rate without duct firing is much better than for the 2x0 simple cycle but has a similar shape, as indicated by this next figure.

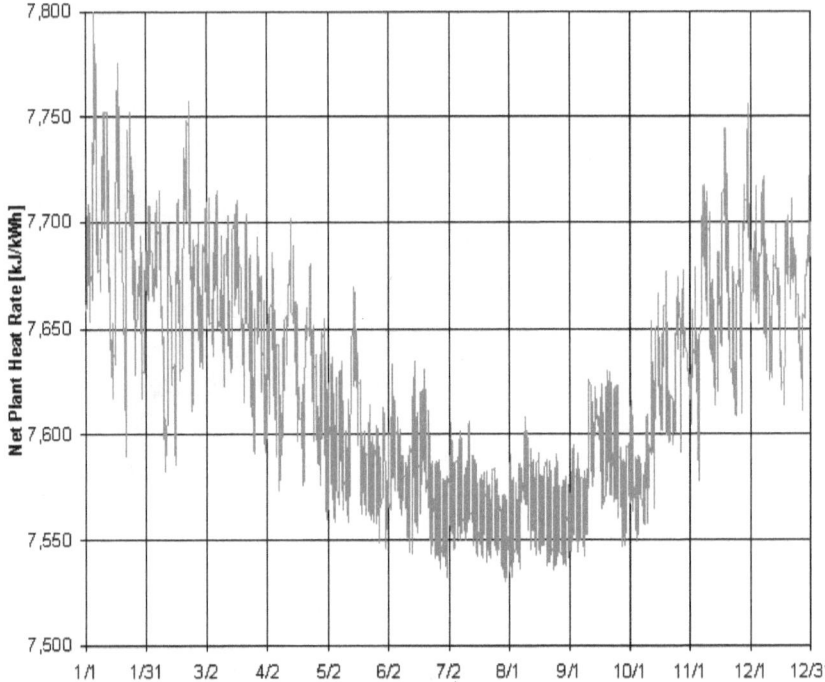

**Figure 59. 2x1 CCPP Heat Rate without Duct Firing**

Seasonal duct firing significantly impacts the heat rate, as shown in this next figure:

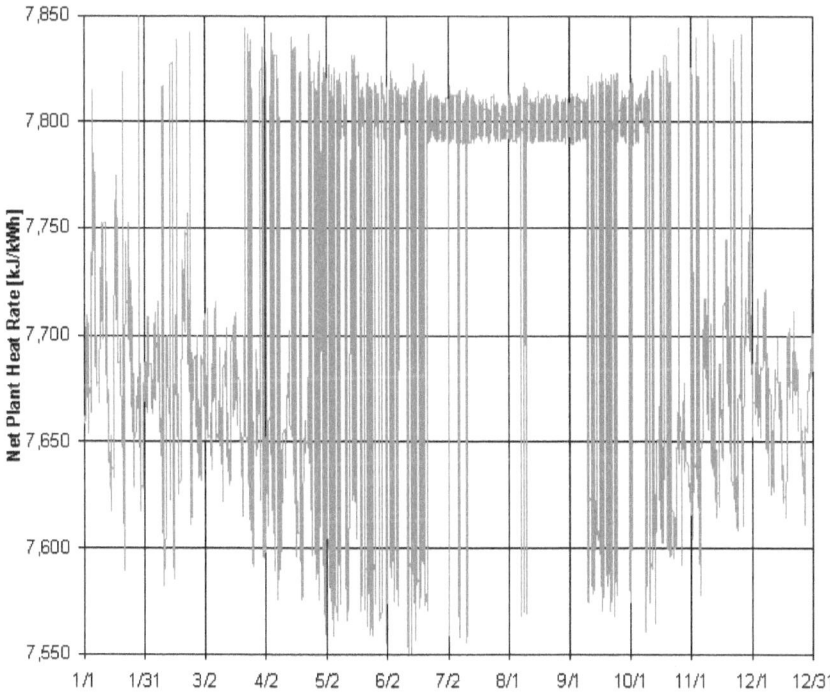

**Figure 60. 2x1 CCPP Heat Rate with Seasonal Duct Firing**

As we have used an average heating value for the fuel, the fuel flow is proportional to the product of the capacity and heat rate

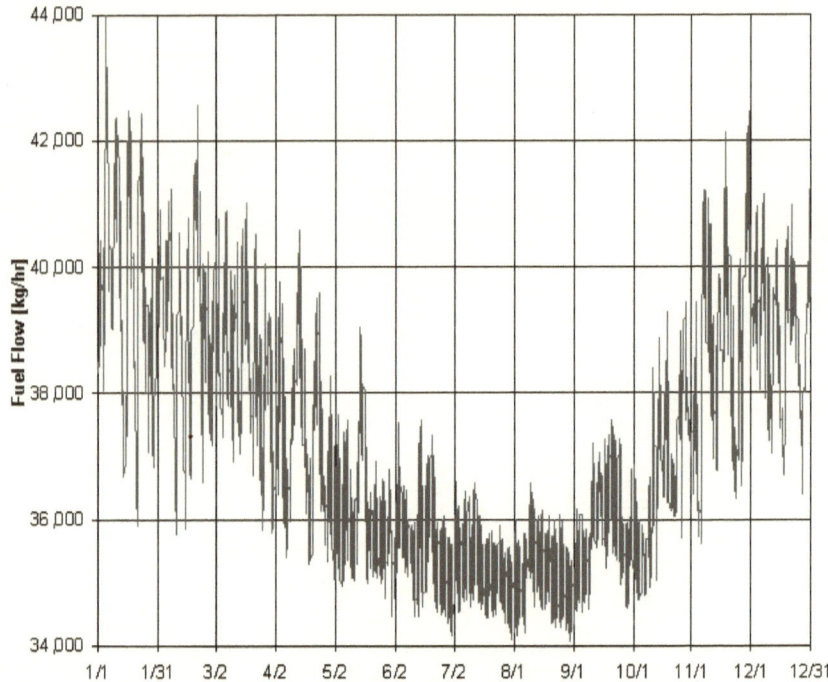

**Figure 61. 2x1 CCPP Fuel Flow without Duct Firing**

Duct firing, of course, impacts the fuel flow, as shown in this last figure:

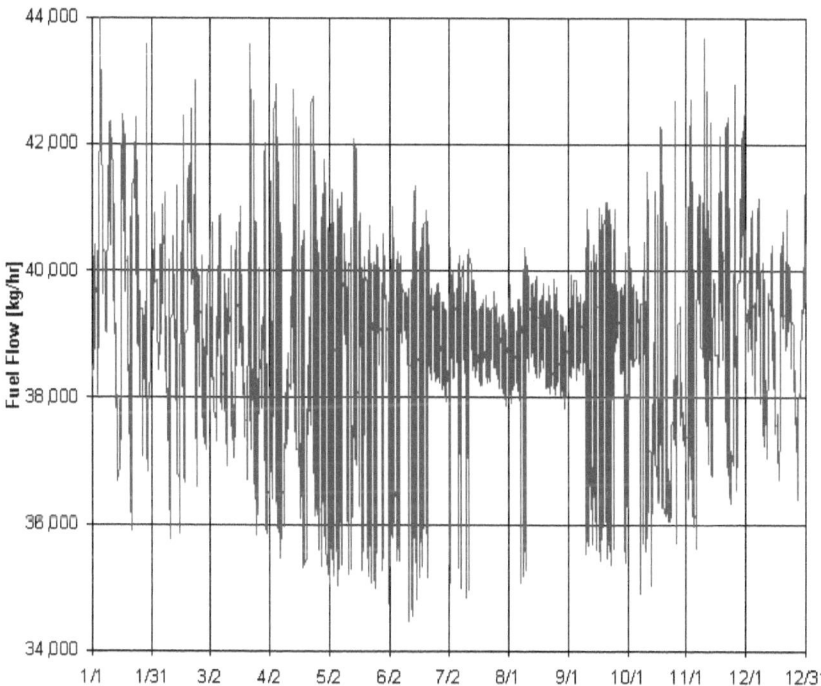

**Figure 62. 2x1 CCPP Fuel Flow with Seasonal Duct Firing**

## Chapter 4. Observed GT Performance

In the preceding chapters we have considered how gas turbines are *expected* to perform, which may not always be the way these are *observed* to perform. We begin with what I suspect is a typical plant. We will first consider nominally full load. As this engine has both inlet conditioning and steam power augmentation, the range of nominally full load operation is broad. We have a full suite of curves from the manufacturer, which will remain anonymous. This suite of curves allows us to calculate the expected performance.

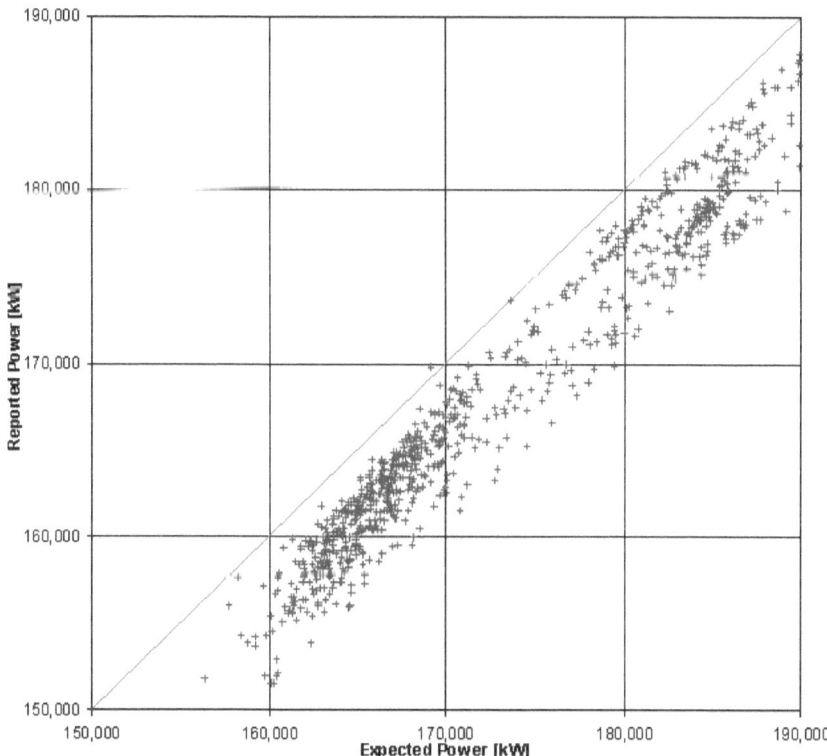

**Figure 63. Reported vs. Expected GT Power**

From this figure we see that the power output is significantly (2.7%) and consistently below that indicated by the manufacturer's curves. This shortfall is why I was brought into to assess the situation. I had no secret formula to fix this disappointing outcome and my comment that, "the engine wasn't too bad and there probably wasn't anything wrong with it," was not what the owner hoped to hear. I was able to adjust their expectation and dispatch, but that's about all I could contribute. It's like buying a sports car. Yours is probably not going to be as good as the one they loaned the car magazine to test drive.

55

This next figure is a comparison of heat input:

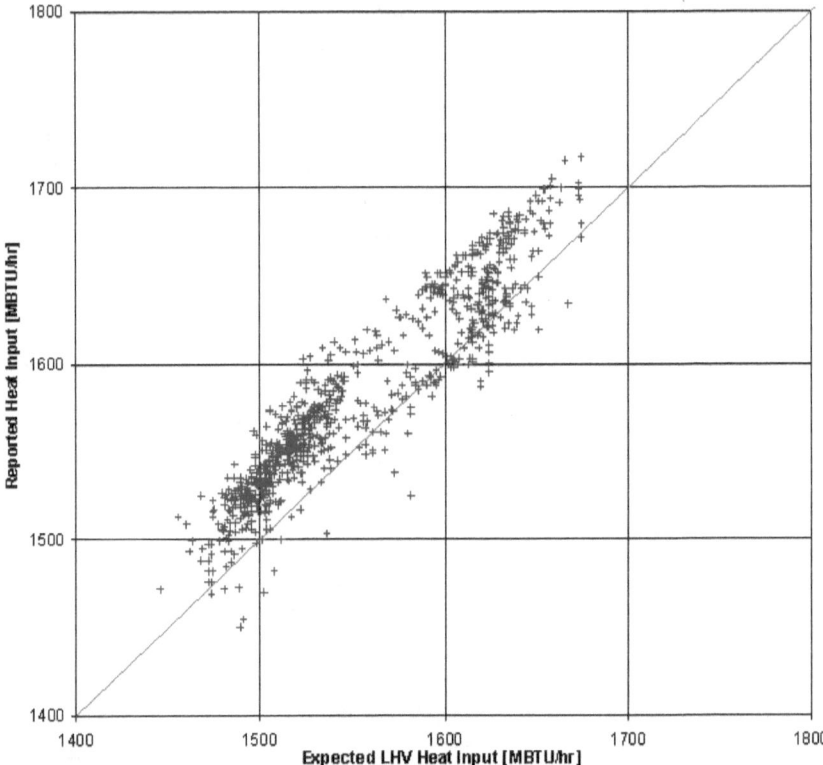

**Figure 64. Reported vs. Expected Heat Input**

Not surprisingly, the heat input is significantly and consistently more (1.8%) than expected, meaning that the performance is doubly disappointing. We will see that in the next figure of heat rate, which is directly proportional to how much it costs to generate electricity. If the owner/operator of such a plant is locked into an agreement or limited by a regulatory commission, this level of performance is not just disappointing, it could mean financial ruin.

This next figure is a comparison of heat rate:

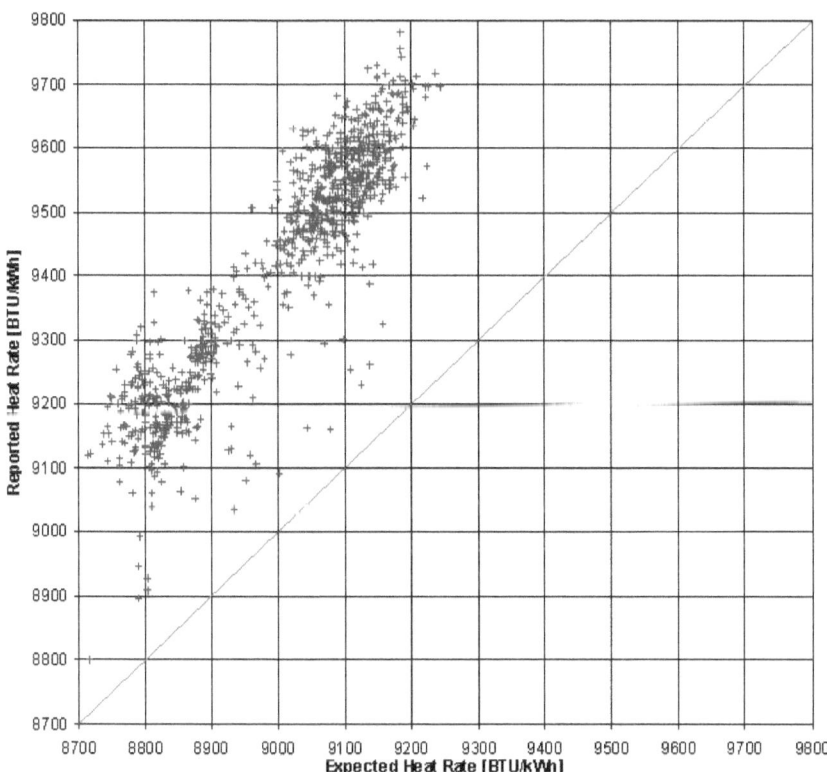

**Figure 65. Reported vs. Expected Heat Rate**

In this figure we see the operating points high (4.4%) and to the left of the manufacturer's expected performance, which is indicated by the diagonal red line. Is it any wonder the owner was distressed over this? This was a small utility in terms of total capacity and relatively new to the field of power production. Had they accumulated decades of experience like some other utilities, the performance of this engine might not have come as such a surprise.

This engine is part of a combined cycle plant and so the GT exhaust is also of value, as it provides heat to the HRSG and ultimately the steam tail. This next figure is a comparison of the GT exhaust flow:

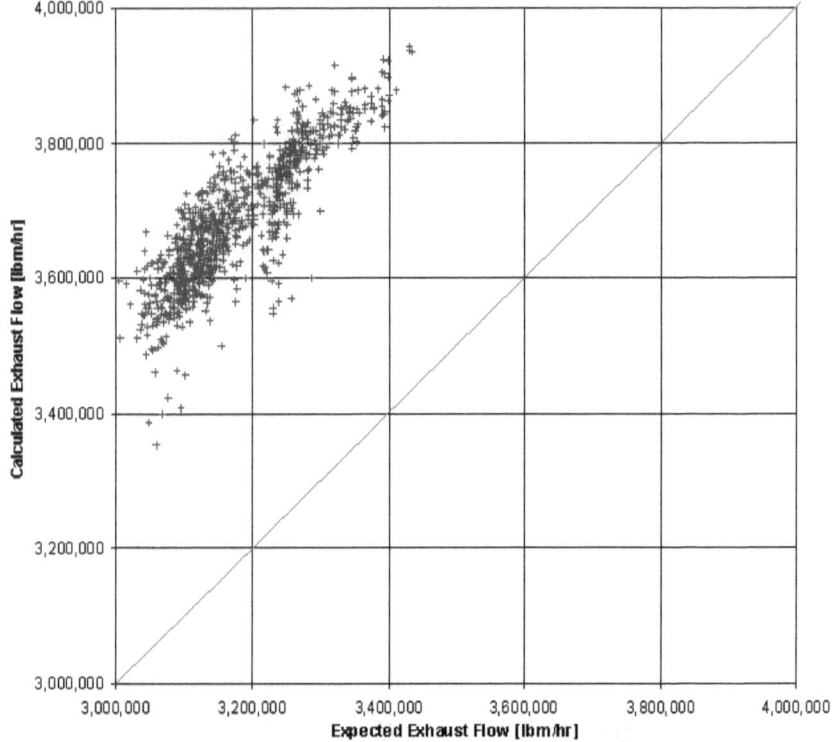

**Figure 66. Reported vs. Expected GT Exhaust Flow**

This result is a little surprising. While the power is lower and heat rate higher, the engine is flowing significantly more (16%!) than the expected air/exhaust. At the very least, this means that there's nothing wrong with the compressor and probably not the expanders either. The first expander drives the compressor and the second drives the generator.

This next figure is a comparison of GT exhaust temperature:

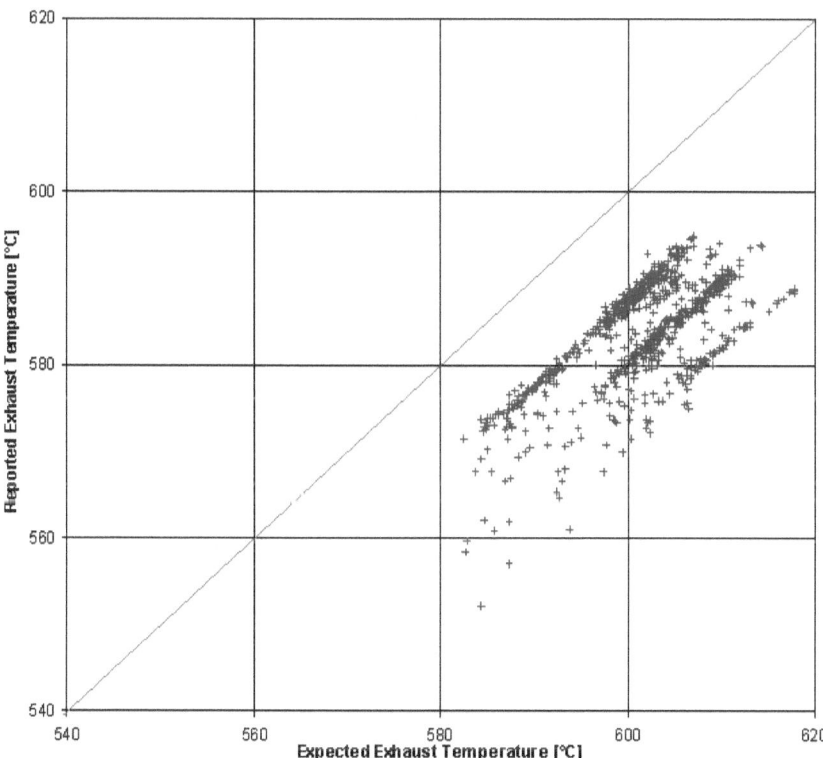

**Figure 67. Reported vs. Expected GT Exhaust Temperature**

*I mixed SI and English units here just to see if you were paying attention.*

Given the heat input, as shown previously, and the higher exhaust flow, the exhaust temperature must be lower (17°C) by conservation of energy (the 1st LoT) and so we see this to be the case in the figure above. At least the energy is going into the HRSG and not directly up a stack into the atmosphere without heat recovery. In this case, raising the exhaust temperature would exceed the emissions limits and so this wasn't an option. The engine was tuned as best as it could be and still meet the regulations. This is something that owner/operators don't always fully consider. An engine that might have stellar performance with only minimal emission restrictions may be crippled by the regulations imposed at a particular site. What might be called *full disclosure* doesn't always happen in the design phase when components and manufacturers are being weighed and trade-offs made. What seems like the best option may not be in the long run and this plant is an excellent example of that.

We see from this next figure that the performance of the engine makes sense, it just doesn't live up to the manufacturer's claims in this particular case, which I reiterate isn't always true. The data to which I am privy is somewhat biased because nobody calls me for help when everything is working as expected. They only call when something is wrong.

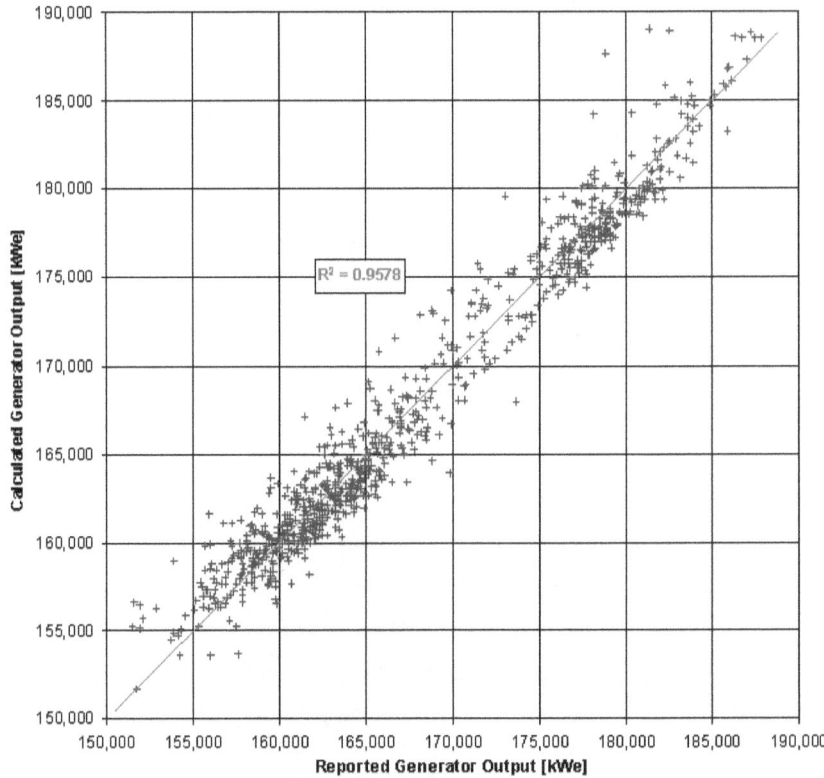

**Figure 68. Results of Power Regression**

This is a complicated configuration, which includes foggers and inlet heating that must be factored into the performance. It also includes steam power augmentation; so at least six parameters come into play for each performance variable.

It is not surprising that the heat input doesn't correlate as tightly as the power, considering the inlet conditioning, steam power augmentation, and variable fuel composition.

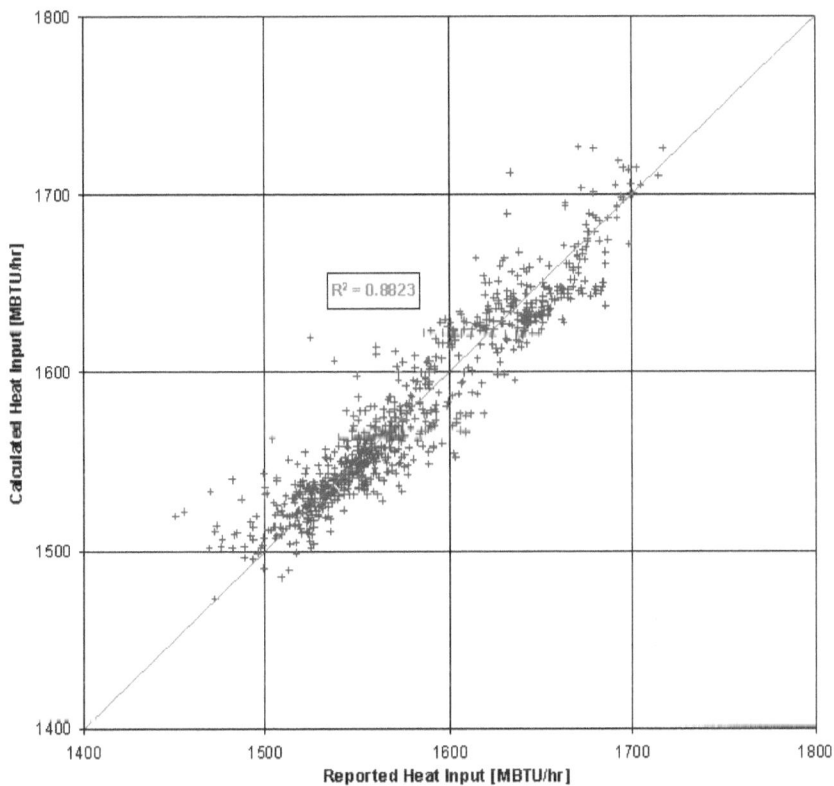

**Figure 69. Results of Heat Input Regression**

Of course, the greatest scatter will be seen in the heat rate, as the ratio of these last two quantities.

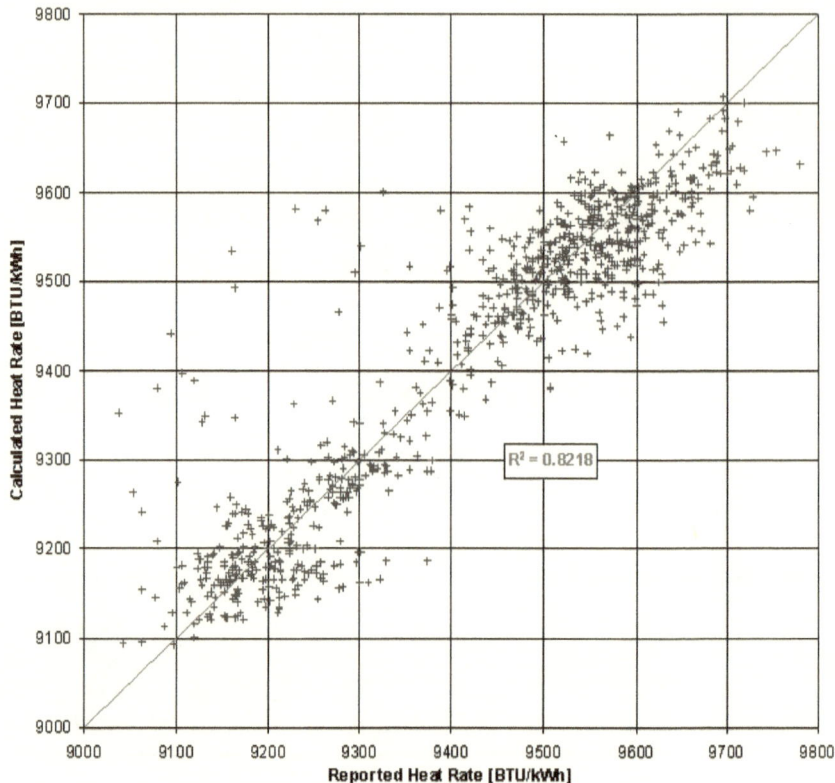

**Figure 70. Results of Heat Rate Regression**

The GT exhaust flow is important to the overall plant performance, as this supplies heat to the HRSG and steam tail. The flow is not directly measured. While this would be possible, it would be quite problematic, given the velocities and temperatures involved. It is customary to calculate the flow using the method outlined in PTC-22.[8] Also see Appendix G for exhaust flow calculation.

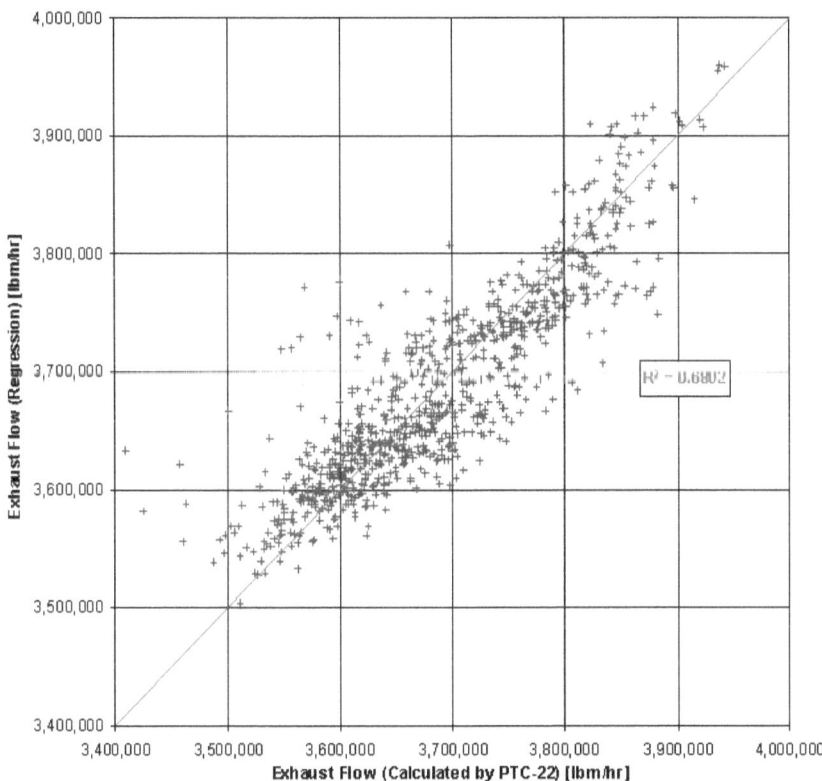

**Figure 71. Results of Exhaust Flow Regression**

There is considerable scatter in the exhaust flow, which is not surprising when considering all of the variables that enter into the calculation and that it's not a direct measurement. Still, the regression is considerably better than the manufacturer's curves.

---

[8] "PTC-22 Performance Test Code on Gas Turbines," American Society of Mechanical Engineers, 2014

The regression coefficient for the exhaust temperature is quite low ($R^2=0.3776$) and the scatter is about the same as the manufacturer's curves, only there is no net bias for the regression (i.e., the data are equally scattered above and below the red diagonal line of exact agreement).

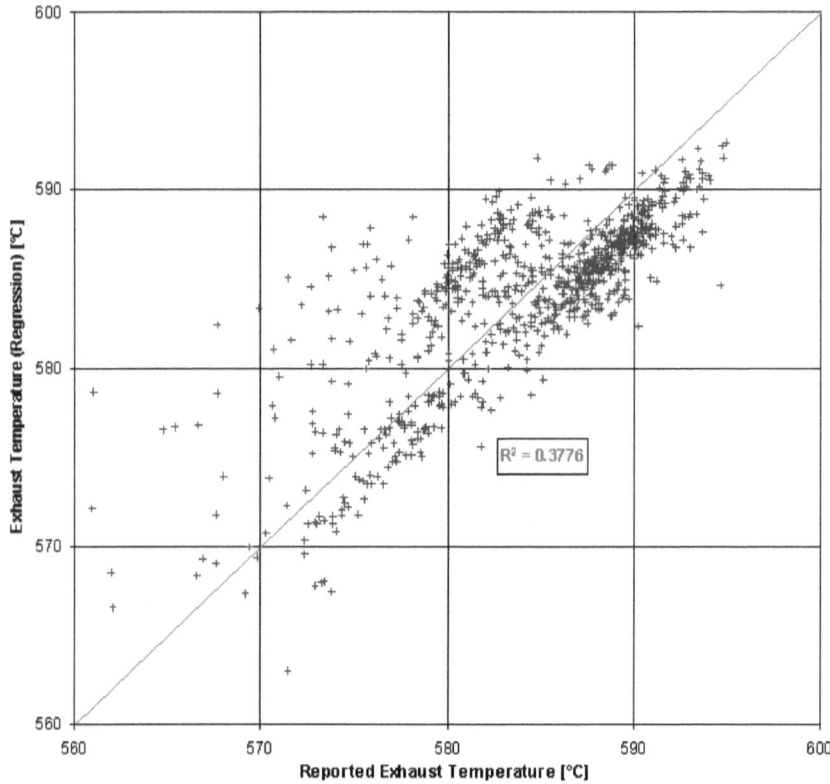

**Figure 72. Results of Exhaust Temperature Regression**

We could go on to show the revised corrections, as before, but there is little point with this particular engine. The discrepancies between observed performance and the manufacturer's curves are not due to the wrong shape but to significant offset or a shift in the performance, which is more indicative of the wrong reference values or baseline performance.

## Chapter 5. Observed CCPP Performance

In the preceding chapters we have considered how combined cycle power plants are *expected* to perform, which may not always be the way these are *observed* to perform. We begin with the most unusual data set I have encountered in the past forty years. It is unusual for two reasons: 1) consistency of operation and 2) reliability of instrumentation. This is so extraordinary; I would like to openly praise those responsible, for this is a remarkable testimony to their diligence. Sadly, the data are proprietary and so the source will not be revealed and adequate details will be obfuscated to preclude identification. The graphics and summary information will be presented herein, but the raw data will not.

The plant from which this data set has been collected is also unusual in that the manufacturer's curves are remarkably inaccurate, as we will see in this first figure. In spite of how bad this looks ($R^2$=-0.0109), do not conclude that every set of manufacturer's curves are bad or even curves from this particular unnamed manufacturer, as I know of many others which are quite adequate. I will not speculate here as to why these particular curves are so bad.

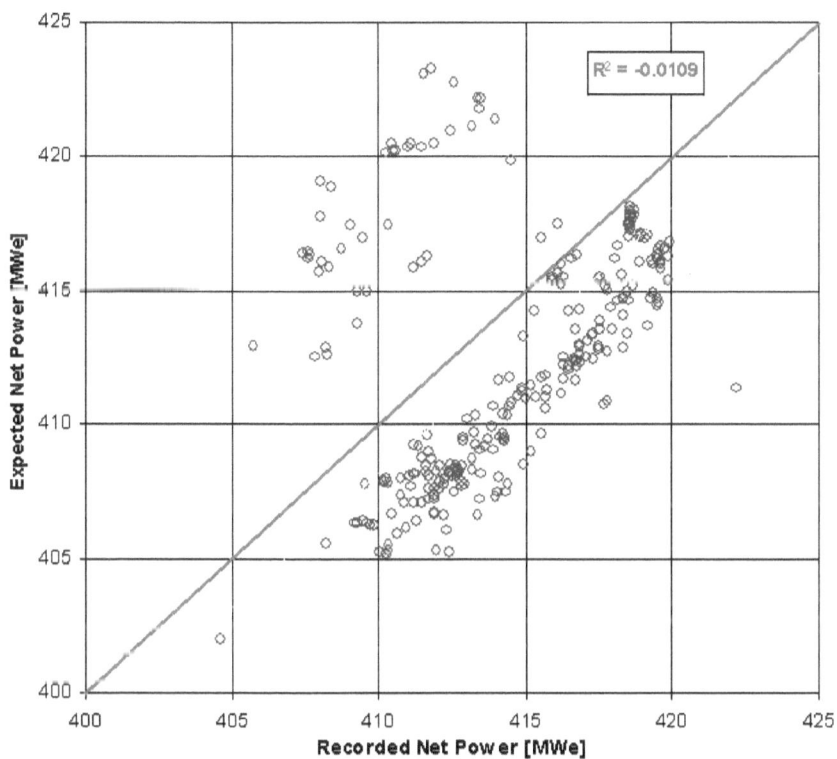

**Figure 73. Recorded vs. Expected Full-Load Net Power Output**

65

When I first discovered this disagreement, I presumed it to be a result of shoddy instrumentation, as this is often the case. Sadly, if a plant continues to run in spite of any particular sensor accurately reporting results, there is little motivation to routinely check and calibrate each and every one. In practice, only a few critical sensors are maintained by the typically over-worked staff, which may consist of a single person tasked with instrumentation. Some companies have instrumentation technicians that go from plant-to-plant and aren't focused on any one particular plant. Which is why this particular data set is so remarkable. Before we introduce accurate curves, let's consider the heat rate.

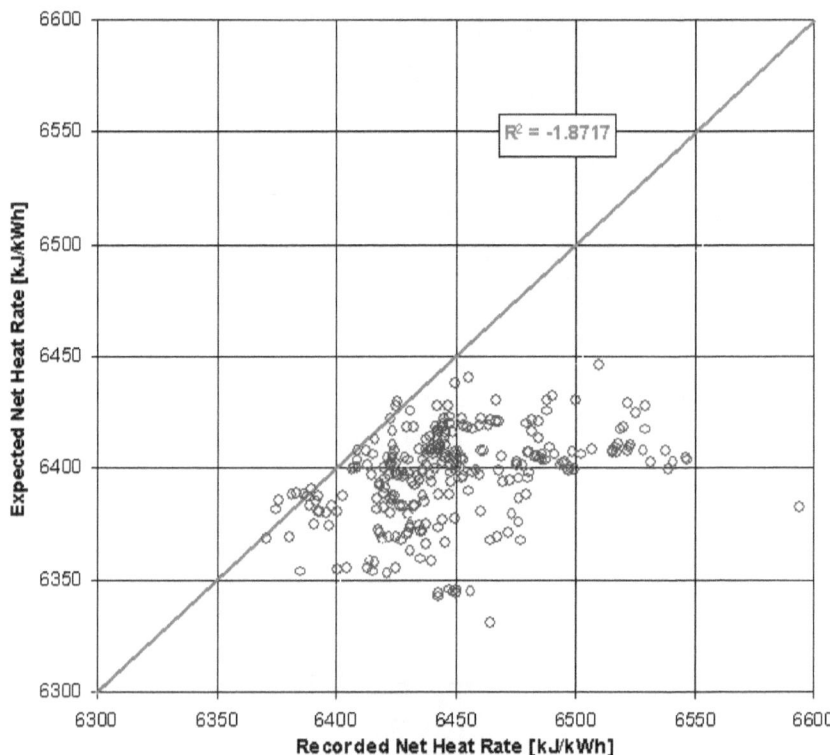

**Figure 74. Recorded vs. Expected Full-Load Net Plant Heat Rate**

And you thought $R^2$=-0.0109 was bad... here we see $R^2$=-1.8717! At least the power was scattered above and below the red diagonal line indicating exact agreement. The heat rate predicted by the manufacturer's curves is consistently below observed (i.e., overly optimistic). Note again that this is at full load, which occurs only about 3% of the time for this plant. What about the rest of the year? How could you accurately dispatch this plant without reliable curves? That's why I was brought in... to remedy the situation.

The last curve we will consider here is the fuel flow, which, surprisingly, is the best of the three ($R^2$=0.4849). At least the correlation coefficient is greater than zero in this case. Again, there is a fairly consistent under-prediction, which is not helpful when making financial decisions.

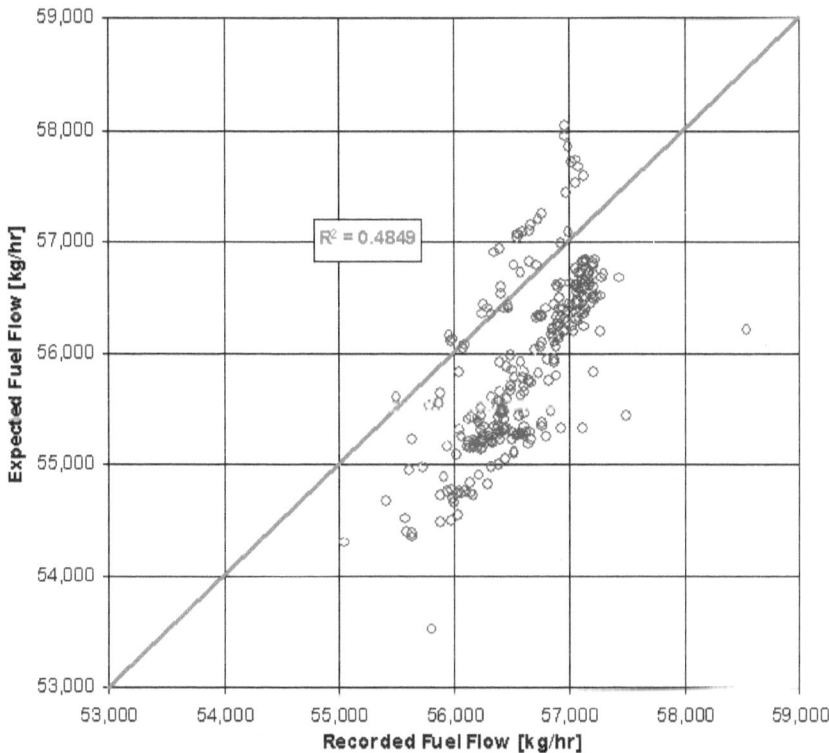

**Figure 75. Recorded vs. Expected Fuel Flow at Full Load**

After a few adjustments to the performance curves, I was able to obtain quite remarkable agreement between calculated and reported performance over a large range of operation. Every one of the curves (compressor inlet temperature, barometric pressure, relative humidity, fuel composition, speed, and aging) required some adjustment. Only the generator loss curves were untouched. It is worth noting here that generator loss curves are typically quite accurate. While it isn't cheap to measure generator losses and efficiency, it is straightforward, which explains why these curves are usually a reliable indication of how the machine will operate.

Results using the revised (customized, site-specific, empirical) performance curves are shown in these next figures, which cover the range of 50% to 100% load and includes 32 times as many data points! First, the net power:

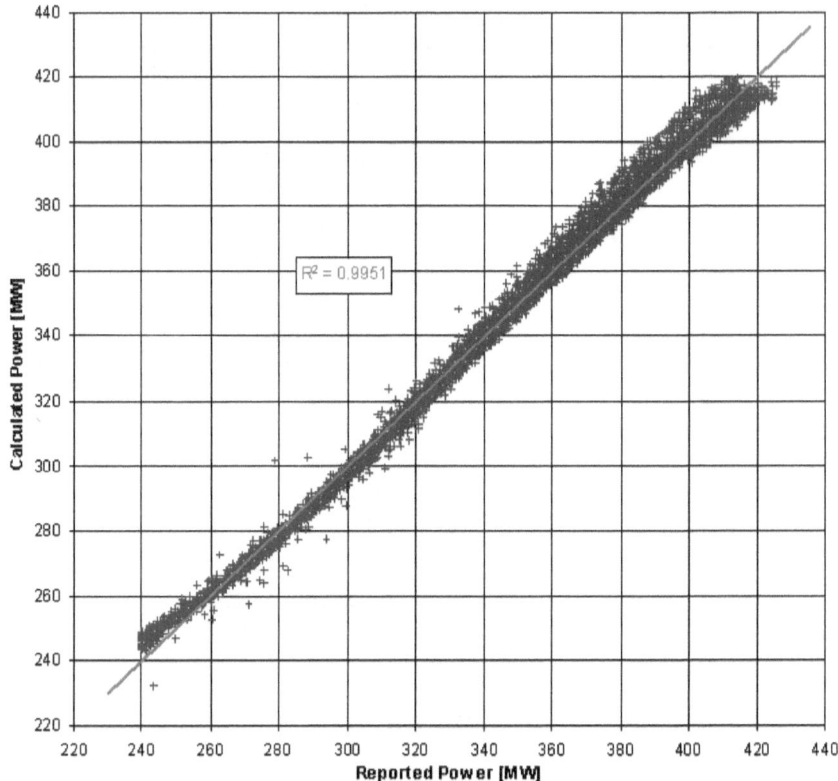

**Figure 76. Revised Net Power Comparison**

The correlation coefficient achieved ($R^2$=0.9951) is nothing short of astounding, which is why I said before that this data set is so remarkable. It is important to note here that this is ALL of the data above 46% load, not just the data points that fit the regression. NONE of the data points (hourly averages) have been removed to make this correlation look better!

Agreement for the heat input is also astounding (R²=0.9949).

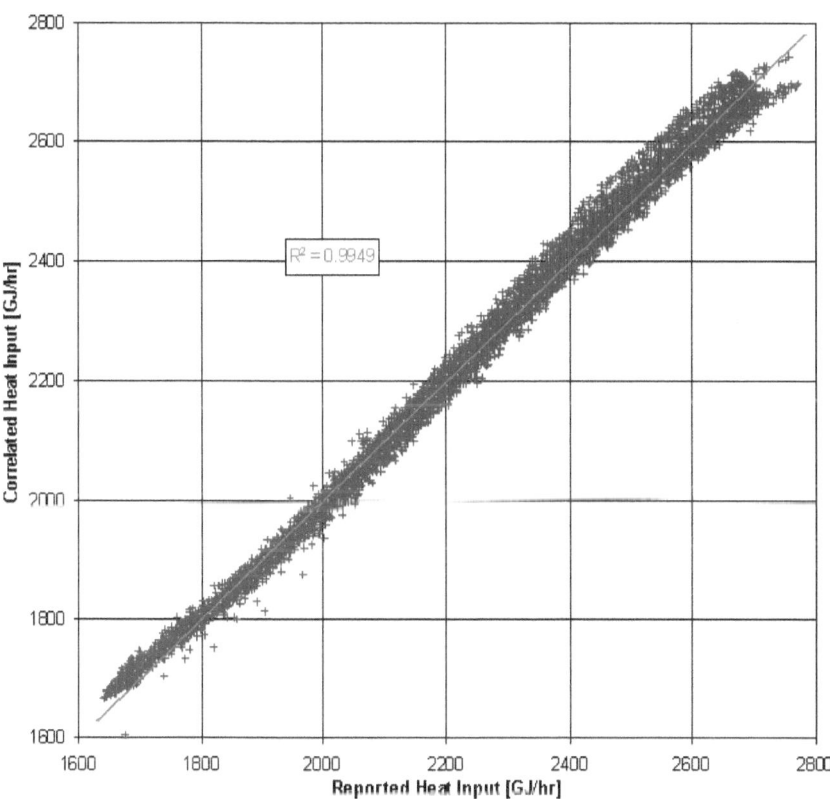

**Figure 77. . Revised Heat Input Comparison**

The comparison of calculated and reported fuel flow is almost exactly the same as the comparison of heat input. After all, they are proportional on a point-by-point basis. They would be exactly proportional over the range if the heating value were constant, but it is not. This means that the corrections accurately capture the changing fuel composition.

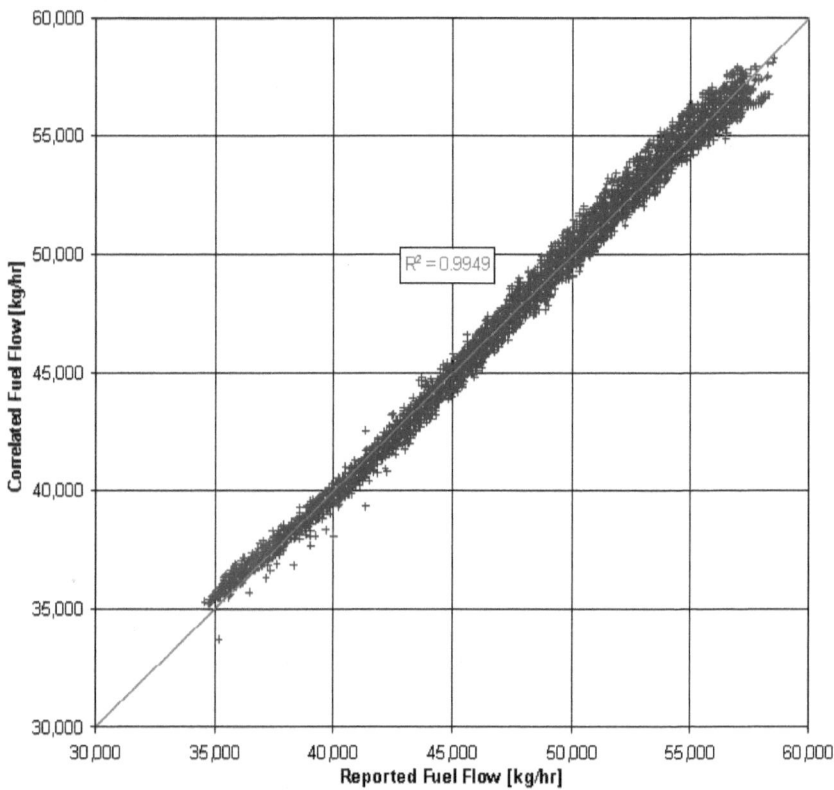

**Figure 78. Revised Fuel Flow Comparison**

We expect heat rate to have more scatter than either power or heat input, as this the ratio of two uncertain quantities. As mentioned in Chapter 1, this is why we correct heat input and then calculate heat rate and not the other way around. This means using Equations 1.1 and 1.2 rather than 1.3 or alpha and beta corrections and not their quotient, f corrections.

This next figure is a comparison of heat rate:

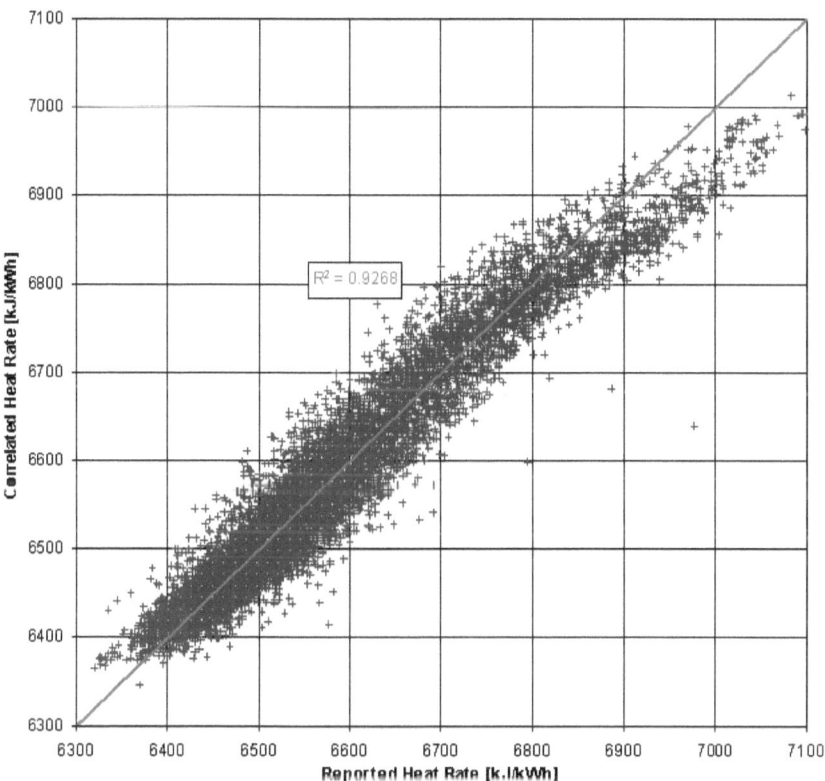

**Figure 79. Revised Heat Rate Comparison**

You will never again see such agreement for this type of data. The correlation coefficient ($R^2=0.9268$) strains the limits of credibility. As a hardcore skeptic, if I hadn't done this myself, I wouldn't have believed it. Consider yourself fortunate, if you can achieve a correlation coefficient of fifty percent ($R^2=0.5$) for heat rate data.

<u>What Went Wrong?</u>

We now consider what went wrong with the manufacturer's curves in this case. I don't know why for certain, although I do have my suspicions. The curves simply don't fit this system. The mistake may have been that simple: the wrong plant. At this point, who knows? We will first consider the impact of compressor inlet temperature, or alpha1 and beta1. Remember that this is a combined cycle plant.

71

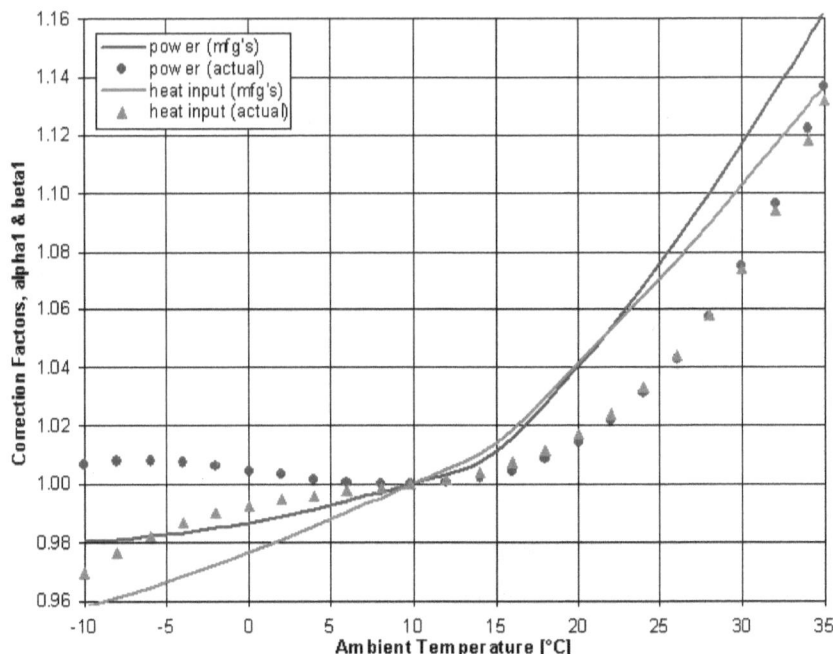

**Figure 80. Temperature Corrections, alpha1 & beta1**

The actual response of the plant to temperature (blue and red dots) is significantly less pronounced than indicated by the manufacturer's curves (blue and red solid lines). The behavior at low temperature (<10°C) is all wrong. This is most likely due to emissions, that is, the control system adjusts the firing to limit emissions at low ambient temperatures, rather then continuing to increase power, as indicated by the blue manufacturer's curve. When a manufacturer offers to provide post-tuning curves, by all means, take them up on the offer, as this is clearly important.

We next consider the impact of barometric pressure. This time we will plot alpha2 and beta2, as f2 is essentially flat.

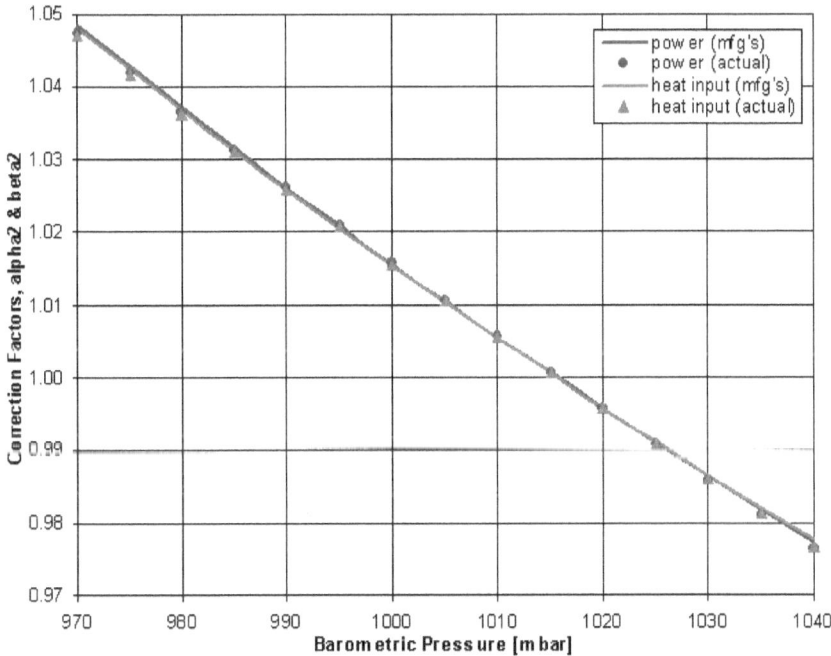

**Figure 81. Barometric Pressure Corrections, alpha2 & beta2**

These are right on, which is why I suspect the curve problem originates in the control system and emissions.

We next consider the impact of moisture. As discussed in Chapter 2, we use absolute humidity rather than relative humidity so that we have single curves instead of a fan of curves.

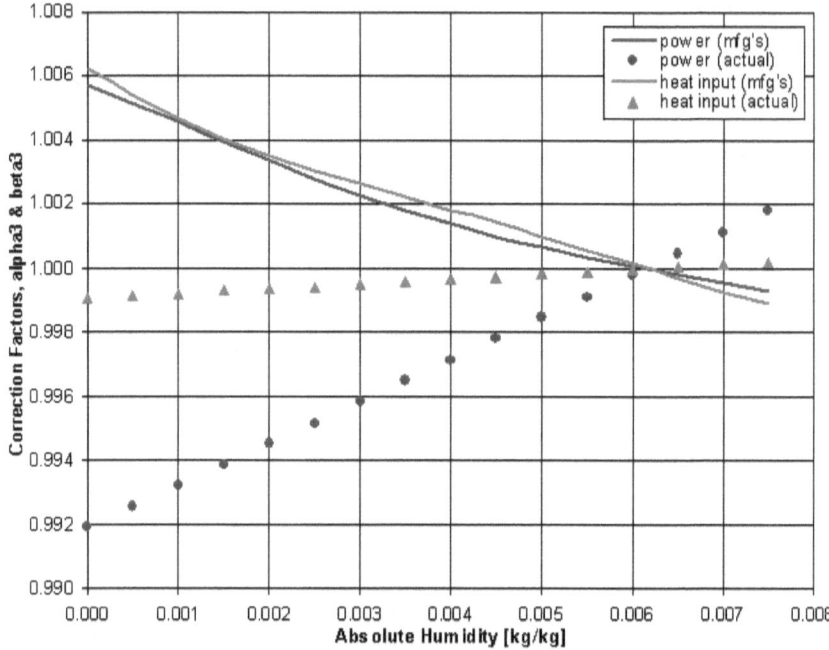

**Figure 82. Humidity Corrections, alpha3 & beta3**

Here we see the manufacturer's curves are opposite of how the plant actually responds to moisture in the air. Not only does water vapor impact the gas turbine operation, but it also impacts the exhaust flow and specific heat, thus energy, entering the HRSG and ultimately the steam tail and net power. We should expect these curves to have a downward slope (correcting power and heat input downward with increasing moisture), but they have an upward slope. This is more evidence that emission control is driving the performance of this plant.

We next consider the impact of fuel composition. While the heating value of the fuel at this plant does change throughout the year, the carbon-to-hydrogen ratio is relatively constant at approximately 3.15, so we will evaluate the correction only on LHV.

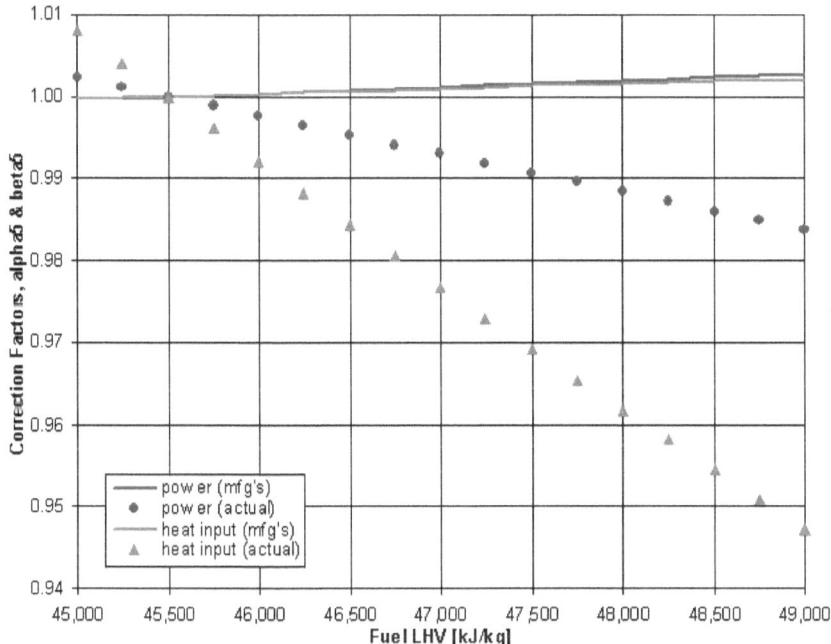

**Figure 83. Fuel Composition Corrections, alpha5 & beta5**

Again, we see the plant responding quite differently than the manufacturer's curves would indicate. This, too, is likely related to emissions.

The final curves we will consider for this plant is the impact of aging.

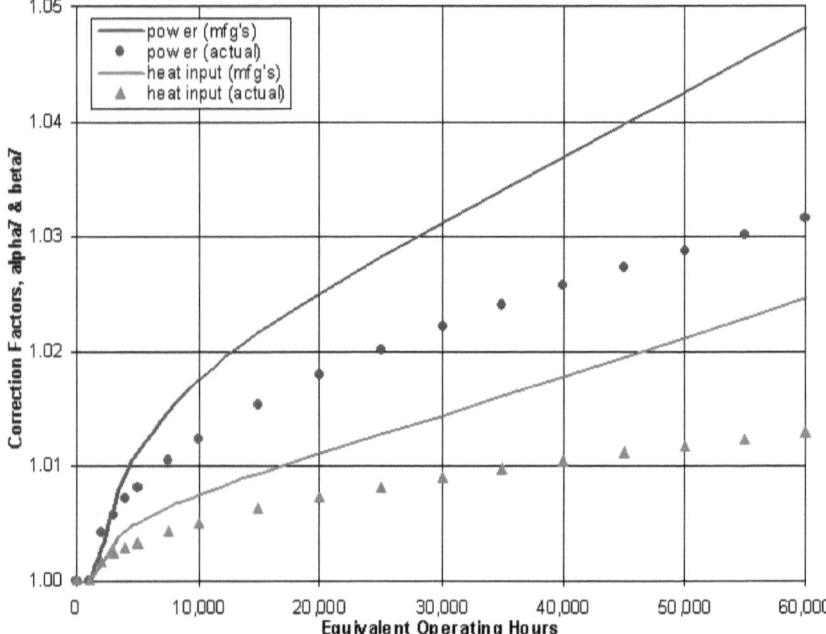

These curves have nothing to do with emissions and everything to do with erring on the pessimistic side by almost exactly 100% or a factor of 2. This is understandable, but something that should neither be overlooked nor taken for granted.

## Appendix A. Thermodynamic Properties of Moist Air

Psychrometrics, or the precise determination of the thermodynamic properties of moist air in the West began with the work of Goff & Gratch, published in a series of papers between 1943 and 1949.[9,10] This work was continued by Hyland & Wexler between the years 1978 and 1983.[11,12,13] Nelson and Sauer further refined with their research between 1999 and 2001.[14] Most recently, Hermann, Kretzschmar, and Gatley have presented a more complete formulation coupled with the latest properties of steam.[15,16]

The basic formulation has been the same since the work of Goff & Gratch. The significant peculiarity of this approach is that the properties are all on a per pound of dry air basis, rather than on a per total (air plus water vapor basis). This facilitates some calculations, which was of greater concern in the 1940s than it is now. Note also that the work of Hyland & Wexler was reported by ASHRAE in English units and so they are continued herein to avoid confusion.

### Saturation Pressure and the Enhancement Factor

It is tempting to simply use the saturation pressure of steam along with Dalton's Law of Partial Pressures[17] to obtain values for the water vapor content in air at saturation, but this isn't accurate. The saturation pressure of steam is for H2O in the vapor state in equilibrium with H2O in the liquid state. This isn't the same as H2O in the vapor state in equilibrium with air in the gaseous state. An

---

[9] Goff, J. A. and Gratch, S., "Thermodynamic Properties of Moist Air," *Heating, Piping & Air Conditioning*, pp. 334-348, 1945.

[10] Goff, J. A. and Gratch, S., "Low-Pressure Properties of Water from -160 to 212 F," ASHVE Trans., pp. 95-122, 1946.

[11] Hyland, R. W., Wexler, A., and Stewart, R., "Thermodynamic Properties of Dry Air, Moist Air and Water and SI Psychrometric Charts," ASHRAE RP-216 and RP-25, 1983.

[12] Hyland, R. W. and Wexler, A., "Formulations for the Thermodynamic Properties of the Saturated Phases of H2O from 173.15 K to 473.15 K," ASHRAE Trans., Vol. 89, pp. 500-519, 1983.

[13] Hyland, R. W. and Wexler, A., "Formulations for the Thermodynamic Properties of Dry Air from 173.15 K to 473.15 K, and of Saturated Moist Air from 173.15 K to 372.15 K, at Pressures to 5 MPa," ASHRAE Trans., Vol. 89, pp. 520-535, 1983.

[14] Nelson, H. F. and Sauer, H. J., "Formulation of High-Temperature Properties for Moist Air," *HVAC&R Research* Vol. 8, pp. 311-334, 2002.

[15] Herrmann, S., Kretzschmar, H.-J., and Gatley, D. P., "Thermodynamic Properties of Real Moist Air, Dry Air, Steam, Water, and Ice," *HVAC&R Research*, 2009.

[16] Herrmann, S., Kretzschmar, H.-J., and Gatley, D. P., "Thermodynamic Properties of Real Moist Air, Dry Air, Steam, Water, and Ice - Final Report," ASHRAE RP-1485, 2009.

[17] Dalton's Law of Partial Pressures states that, in a mixture of non-reacting gases, the total pressure is equal to the sum of the partial pressures exerted by each of the constituents and these individual contributions to the whole are each proportional to the mole fraction of that component.

*enhancement factor*, f, is introduced to account for this difference. The enhancement factor is equal to the partial pressure of water vapor that should produce the observed content divided by the saturation pressure of steam at that same temperature. The values of f are close to unity.

Herrmann, Kretzschmar, and Gatley, present a very complicated equation in Section 3.4.2.1 of their report for ln(f) in terms of second and third virial coefficients[18], explaining that this arises from Henry's Law.[19] While this is interesting and may facilitate the calculation of f at elevated pressures without necessitating experiments, it is immaterial. All that is needed is to measure and tabulate the water content of air. An explanation as to why it is what it is, is not essential. This is the approach that Goff & Gratch and Hyland & Wexler took. The following table of f can be found in any edition of the *ASHRAE Handbook of Fundamentals*.

Enhancement Factor, f

| T°F | Pressure [in.Hg] | | | | | |
|---|---|---|---|---|---|---|
| | 10 | 15 | 20 | 25 | 30 | 32 |
| 0 | 1.0016 | 1.0025 | 1.0033 | 1.0040 | 1.0047 | 1.0051 |
| 20 | 1.0016 | 1.0024 | 1.0032 | 1.0039 | 1.0045 | 1.0048 |
| 40 | 1.0018 | 1.0025 | 1.0032 | 1.0038 | 1.0044 | 1.0047 |
| 60 | 1.0020 | 1.0026 | 1.0033 | 1.0039 | 1.0044 | 1.0047 |
| 80 | 1.0023 | 1.0029 | 1.0036 | 1.0041 | 1.0046 | 1.0049 |
| 100 | 1.0027 | 1.0033 | 1.0040 | 1.0045 | 1.0050 | 1.0053 |
| 120 | 1.0031 | 1.0037 | 1.0044 | 1.0050 | 1.0055 | 1.0057 |
| 140 | | 1.0041 | 1.0048 | 1.0054 | 1.0059 | 1.0063 |

The variation of f with temperature at 1 atm. is shown in the following. For the purpose of calculations, the humidity ratio, W, is needed. This can be calculated from f, Psat, and the molecular weights by Equation A.1:

$$W = \left( \frac{MW_{H2O}}{MW_{AIR}} \right) \left( \frac{fP_{SAT}}{P_{BARO} - fP_{SAT}} \right) \tag{A.1}$$

The molecular weight of water is 18.01528 and of air is 28.9645. Pbaro is the barometric pressure. Psat is the saturation pressure of steam in the same units as the barometric pressure. The denominator in Equation A.1 becomes zero when fPsat=Pbaro, which is why this formulation can't be used at elevated temperatures.

---

[18] The virial expansion of the equation of state was first proposed by Kamerlingh Onnes in 1901. It forms the basis for many developments in thermodynamics related to the properties of fluids. It is... $Z=PV/RT=1+B\rho+C\rho^2+...$

[19] Henry's Law states that, at a constant temperature, the amount of a gas that will dissolve in a liquid is directly proportional to the partial pressure of that gas in equilibrium with that liquid. William Henry 1803.

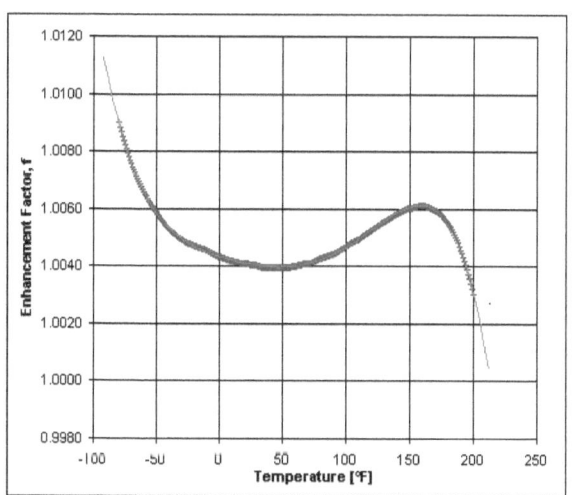

The variation of Psat and W with temperature at 1 atmosphere barometric pressure is shown in this next figure:

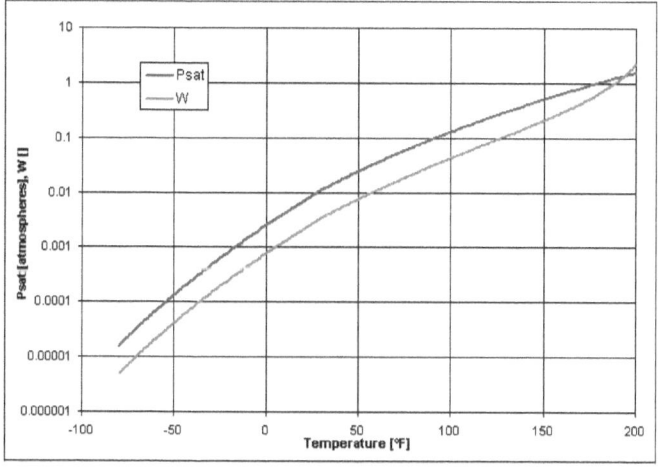

<u>Enthalpy and Entropy</u>

The enthalpy of moist air is also calculated on a per unit mass of dry air basis. Over the range of interest (-80°F to 212°F/-62°C to 100°C), the specific heat, Cp, of air varies so little that 0.24 BTU/lbm/°F (1 kJ/kg/°C) is an adequate representation. The enthalpy of water vapor varies linearly over this range so that the following equation is adequate:

$$h_G = 1061 + 0.444\,T \tag{A.2}$$

Temperature is in degrees Fahrenheit and enthalpy is in BTU/lbm. Conversion to degrees Celsius and kJ/kg is trivial. It is very important to stress here that the appropriate enthalpy of water vapor is $h_G$ and NOT $h_{FG}$ (that is, the enthalpy of the saturated vapor, not the latent heat of vaporization or the difference between the vapor and liquid enthalpies). Although many will insist the latter is correct-this statement even appears in print-it is *not* true. Consider the case of steam at the critical point mixing with dry air. At the critical point $h_{FG}=0$, yet considerable heat accompanies steam at the critical point.

The full equation for enthalpy is:

$$h = h_A + W h_G = 0.24\,T + W(1061 + 0.444\,T) \tag{A.3}$$

The entropy is a little more complicated, because of the partial pressures:

$$s_A = 0.24 \ln\left(\frac{T + 469.67}{469.67}\right) \tag{A.4}$$

$$s_G = 2.29688 - 0.003692687\,T + 0.0000055T^2 \tag{A.5}$$

$$s = s_A + W s_G - R \ln\left(\frac{P}{14.696}\right) \tag{A.6}$$

### Range of Applicability

For the most part these properties are used for atmospheric processes, which is not a problem. These properties are NOT valid near or above the boiling point of water or at high pressure; therefore, they must NOT be used beyond the inlet of a gas turbine. This preclusion extends to the compressor, combustor, and expander. It also applies to the heat recovery steam generator. Should you need the properties of moist air at conditions other than atmospheric, use the NASA Glen Report.[20]

### Excel AddIn

If you need it, you can get a free Excel AddIn from my web site listed in the Forward. Search for *psychrometrics*.

https://dudleybenton.altervista.org/software/index.html

---

[20] McBride, B. J., Zehe, M. J., Gordon, S., "NASA Glenn Coefficients for Calculating Thermodynamic Properties of Individual Species," NASA Report No. 211556, 2002.

## Appendix B. Higher vs. Lower Heating Value

I have mentioned several times that higher heating value is a meaningless quantity and here's why... The difference between higher and lower heating value of a fuel containing hydrogen is whether the resulting water ($H_2O$) is in the liquid or vapor phase. The difference is the latent heat of vaporization. In the nineteenth century when industries first began buying coal in large quantities, there was concern over the value of different coals. Russian chemist, Dimitri Mendeleev, who started development of the periodic table, worked to quantify the fair price of coal based on the heating value. French chemist, Paul Vieille, who invented smokeless powder and also the bomb calorimeter. furthered the process.

If the final temperature of the calorimeter is low enough so that the water formed is in the liquid state (often the case), then what you have measured is the higher heating value. If the final temperature of the bomb calorimeter is high enough that the water formed is in the vapor state, then what you have measured is the lower heating value. We don't use bomb calorimeters to measure the heating value of natural gas. We use a gas chromatograph. Rest assured that the exhaust leaving the stack of a modern combined cycle power plant is hot enough that water is in the vapor state and lower heating value is the appropriate measure.

The motivation for using higher heating value in modern times is a holdover from the nineteenth and early twentieth centuries when coal purchasers feared they were being cheated by coal suppliers. Higher heating value has no legitimate connection to natural gas. Consider this... GateCycle™ (the General Electric heat balance software developed by Michael Erbes, Milt Venetos, Peter Pechtl, and others) doesn't even use HHV for anything. It only uses LHV. When you build such a plant in Antarctica that blows rain or snow out the stack, feel free to consider higher heating value. Until then, use only LHV.

## Appendix C. C/H vs. H/C Ratio

Some gas turbine manufacturers use C/H, while others use H/C to characterize fuel composition. This can be confusing, especially considering they both typically range in value from 3 to 4 for natural gas. C/H is a mass ratio and H/C is a ratio of moles. The molecular weight of carbon is 12, while the molecular weight of hydrogen is 1. The simple paraffin series (methane, ethane, propane, butane, pentane, hexane, heptane, octane, nonane, and decane) have molecular formulas $C_N H_{2N+2}$ ($CH_4$, $C_2H_6$, $C_3H_8$, etc.). The numbers just happen to work out to approximately the same value. Make sure you know which one the manufacturer means and make sure you're using the correct formula. This is another reason why it's good to have a reference fuel composition associated with each installation. The calculation used by General Electric is:

```
H_C=(4*xCH4+6*xC2+8*xC3+10*(xiC4+xnC4)+12*(xiC5+xnC5)
+14*xC6+16*xC7+2*xH2)/(xCH4+2*xC2+3*xC3+4*(xiC4+xnC4)
+5*(xiC5+xnC5)+6*xC6+7*xC7)
```

where xCH4 is the mole fraction of methane, xnC4 is the mole fraction of n-butane, xiC4 is the mole fraction of isobutane, etc. up to heptane.

Note that the Gas Producers' Association Standard (GPA-2145) uses the mass ratios and is considerably more tedious, as natural gas composition is most often given in mole (or volume, assumed to be equivalent to mole) ratios so that the mass fractions must first be calculated.

## Appendix D. Generator Curves

Generator curves are an important part of a performance package. There are several types of curves and the practicing engineer must know what these are and how to use them. One of the more common graphs is shown below:

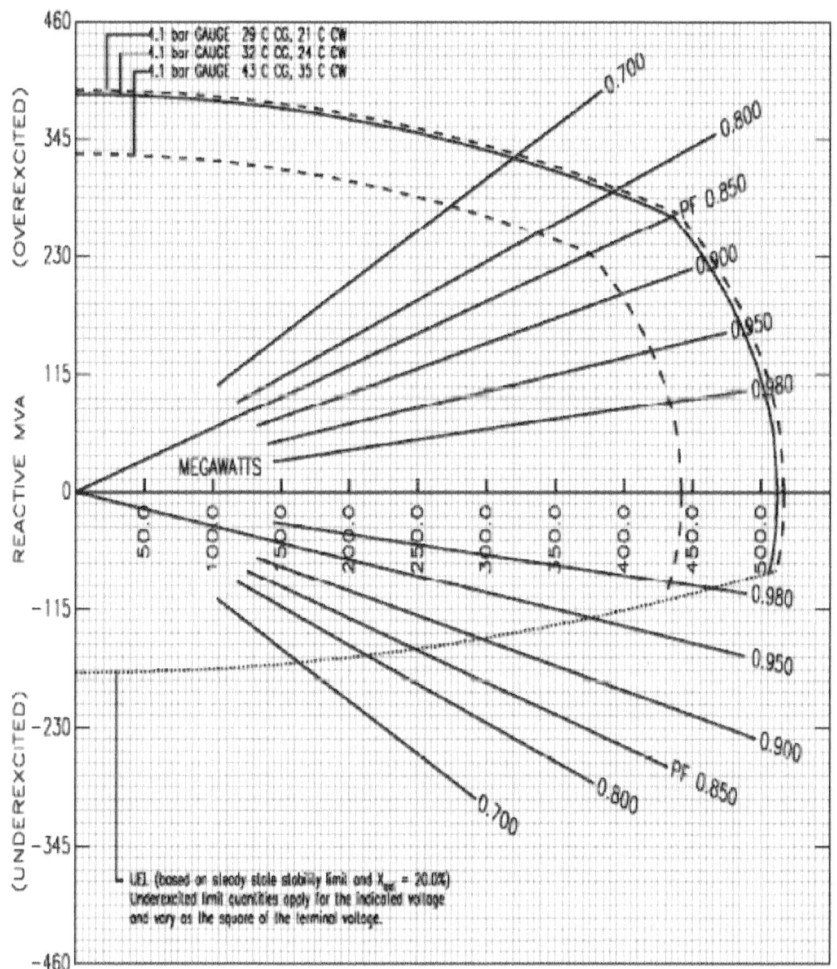

**Figure 84. Typical Reactive Capability Curves**

While these might be of some interest for sizing or diagnostics, they are almost useless for evaluating performance. It is possible to infer efficiency and losses from this information but quite tedious and not worth describing the convoluted process here.

This next figure is totally useless when it comes to performance:

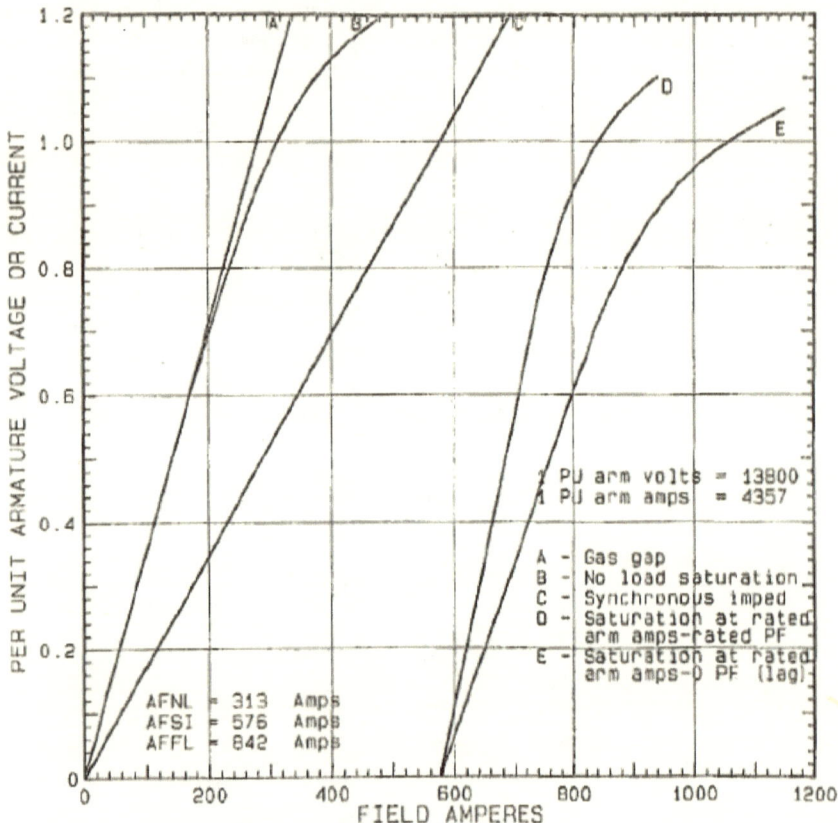

Figure 85. Typical Synchronous Impedance Curves

This next figure is not ideal but adequate. These curves show efficiency as a function of net power for several values of power factor.

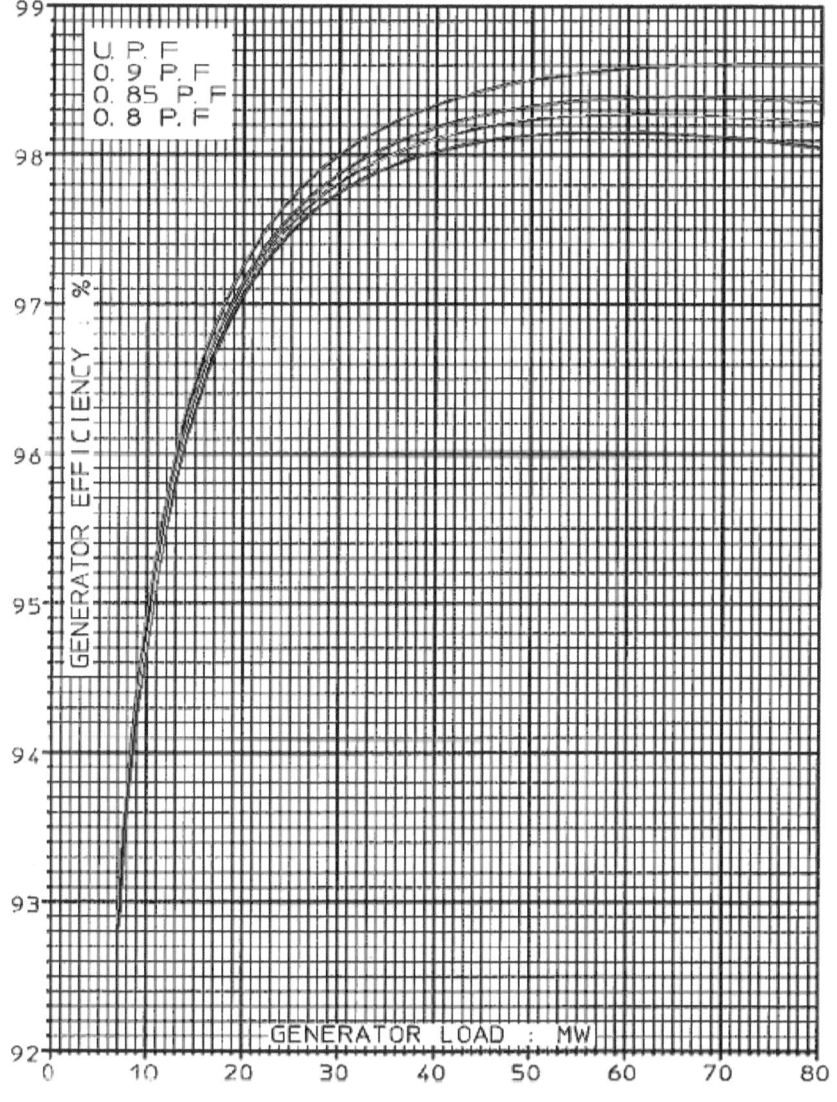

**Figure 86. Typical Efficiency Curves**

While these do provide the necessary information, they are very steep and close together at low power and very flat at high power. This makes them difficult to accurately digitize. No simple polynomial ($y=a+bx+cx^2+dx^3$) will reproduce this shape; so that is also a problem with these curves.

The curves below are what you want to specifically request. They are easily digitized and a simple polynomial will fit quite well.

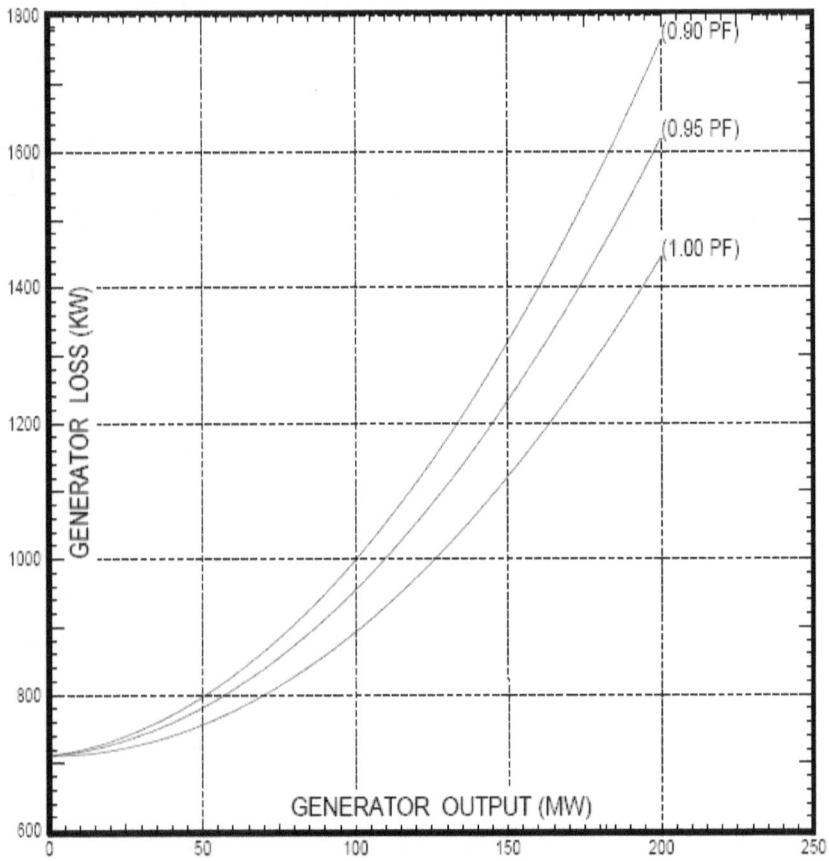

**Figure 87. Typical Generator Loss Curves**

Tools are available for digitizing curves such as these. You will find a free tool on my web site at the link listed in the Forward. This tool is particularly convenient. Simply use the snapshot tool to copy the figure from the PDF to the clipboard and then open the digitizer. It will automatically paste the image from the clipboard. Right click on the four corners and define the scale and transformation and then left click along each of the curves.

For generators and transformers, the multiple curves at different power factors will collapse to a single curve if you divide by the power factor raised to some exponent between 1.25 and 2.5. A simple regression can be obtained using the Excel function LINEST and the exponent can be optimized using the Excel

Solver. You will find this data and regression process illustrated in the spreadsheet generator_loss_curves.xls

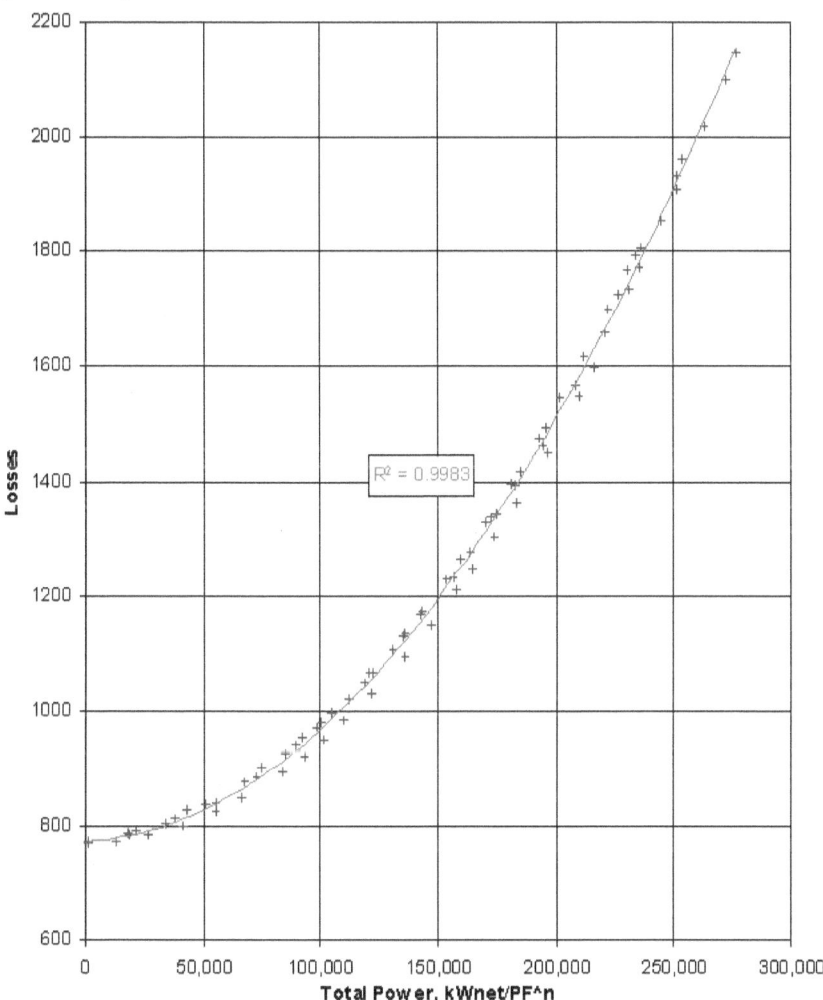

**Figure 88. Multiple Loss Curves Collapse to Single Curve**

| LINEST | | | |
|---|---|---|---|
| -3.5E-15 | 1.87E-08 | 9.11E-05 | 774.1076 |
| 4.19E-15 | 1.71E-09 | 0.000196 | 6.096061 |
| 0.998547 | 15.10124 | #N/A | #N/A |
| 16718.52 | 73 | #N/A | #N/A |
| 11437839 | 16647.45 | #N/A | #N/A |

89

The exponent (in this case 1.5048) can be optimized to minimize the overall residual or maximize the F-value. See Excel help on LINEST for details of which cells contain these values. The final results for this set of curves, which is a little different from the ones shown above, are:

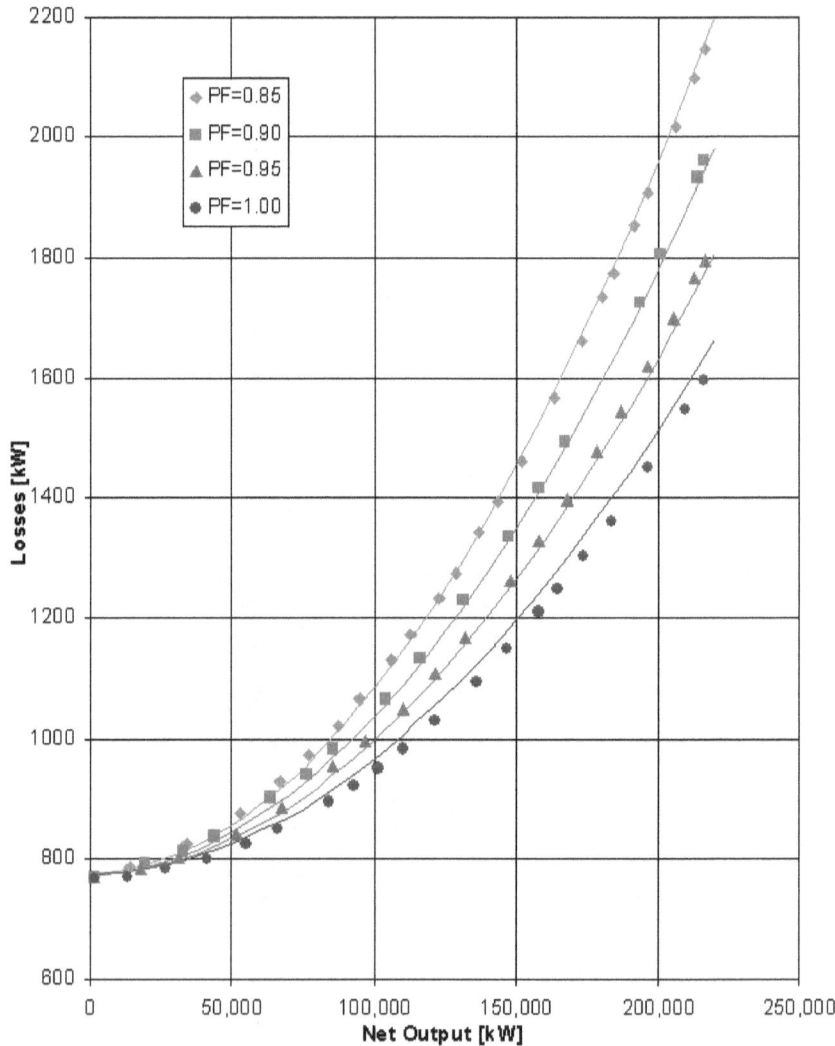

**Figure 89. Typical Generator Loss Curve Regression Results**

## Appendix E. Multivariate Regression

Single variable regression (y=a+bx+cx²) and multivariate regression (z=a+bx+cy+dx²+exy+fy²) are quite useful when developing approximations such as performance curves. The free tool (CurveFit) available on my web site listed in the Forward can be a real time-saver. Copy the data as columns (no commas in the numbers) to the clipboard and then push the button to paste them into the program. For example a typical relative humidity correction can be obtained as follows:

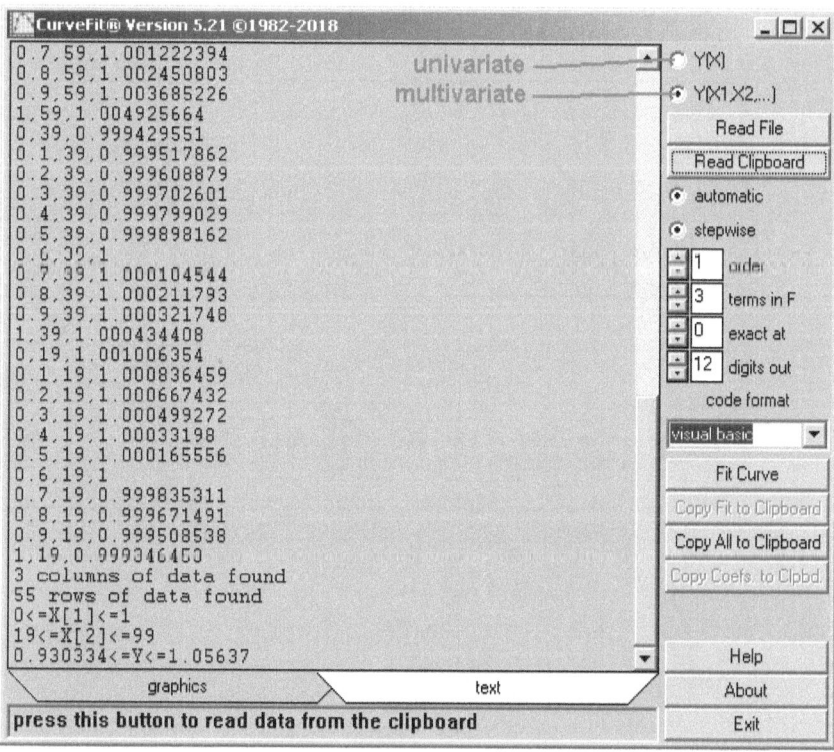

There are several options available, including manual and automatic selection of terms. The output can be formatted for C, FORTRAN, or VBA (Excel macros).

Make sure that there are no commas within numbers, only between columns. Data can be separated by commas or spaces and can be read from the clipboard or from a file.

CurveFit will accept up to 9 columns of data. The last column is the dependent variable and the previous ones are the independent. Do not include extraneous variables, as there is no way to exclude data columns and they must appear in this order.

91

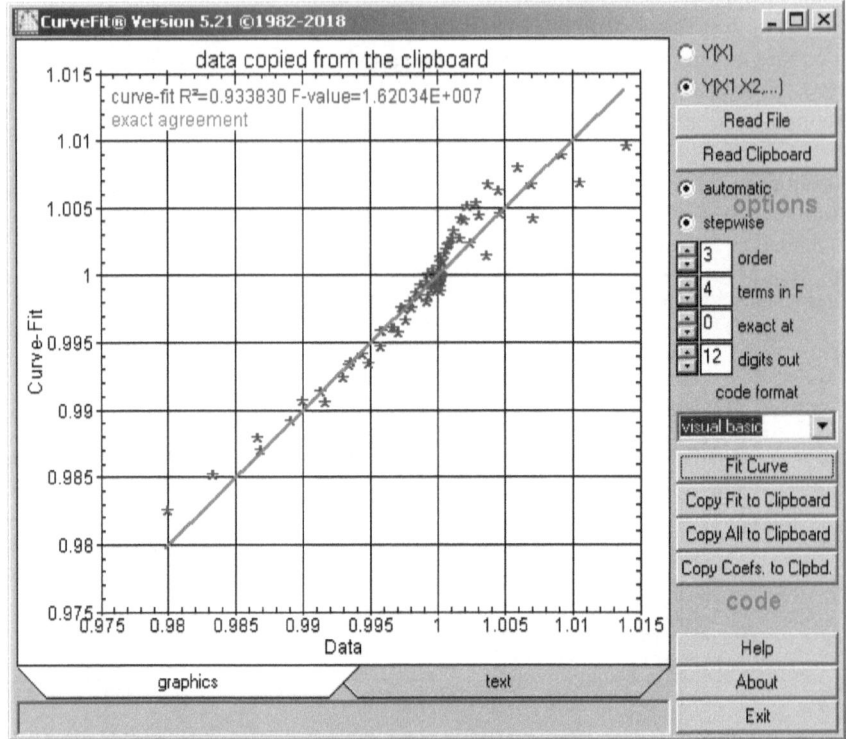

The results are displayed graphically, including the regression coefficient ($R^2$) and F-Value. Below is a sample of what the generated code looks like:

```
Function Y(X1 As Double,X2 As Double)As Double
    Y=9.98956151506E-001
    Y=Y+2.70911282331E-006*X1*X2^2
    Y=Y-1.65917512537E-008*X2^3
End Function
```

## Appendix F. Risk of Icing

While not strictly a performance curve, the risk of icing at the inlet of a gas turbine or inside the compressor in some locations is significant and the potential for damage can be devastating. Velocities at the inlet of a gas turbine are quite high plus the pressure drops below ambient so that the potential for ice formation is much higher than ambient. This risk can be illustrated graphically to include temperature, moisture (in terms of dew-point), and velocity (in terms of mach number).

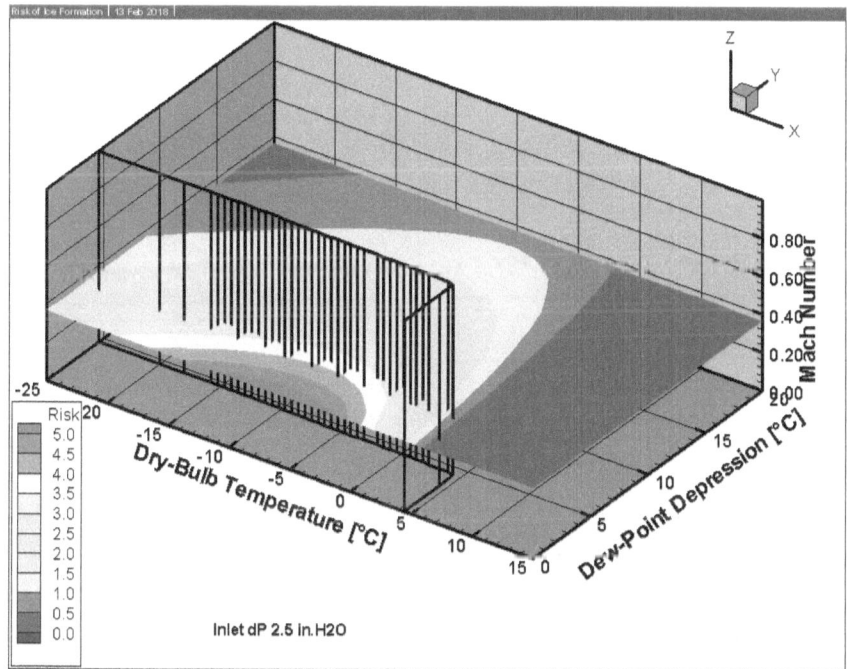

The icing risk factor is assigned a relative index from 0 (none) to 5 (certain). The black *fence* shown in the figure above is the manufacturer's recommended operational boundary (i.e., don't operate inside the fence). The calculation includes compressive heating. The multivariate shape of the operational boundary to avoid is shown in these next figures.

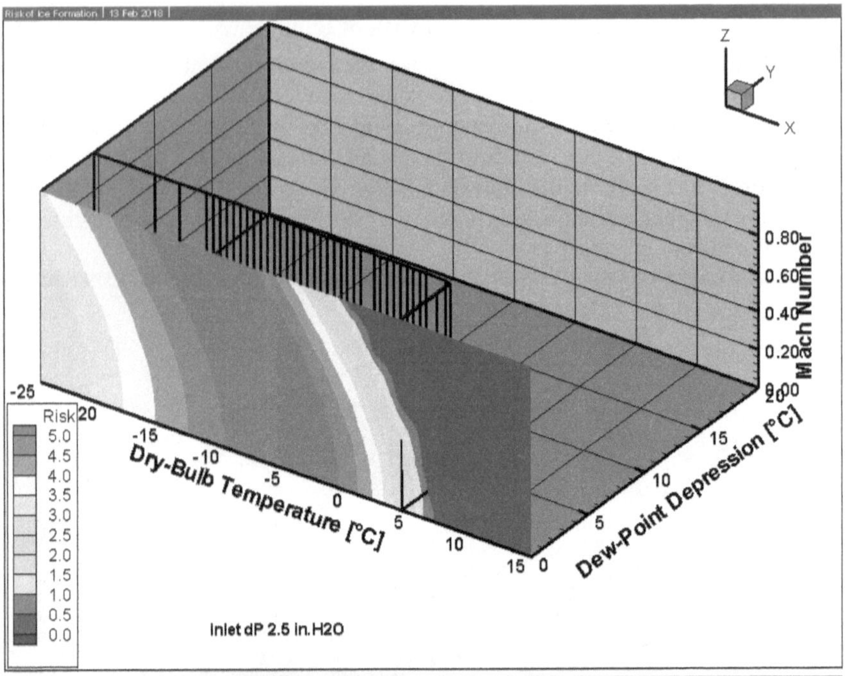
Risk of Ice Formation | 13 Feb 2018
Z
Y
X
0.80
0.60
0.40
0.20
0.00
Mach Number
-25
20
15
10
5
0
Dew-Point Depression [°C]
-15
-10
-5
0
5
10
15
Dry-Bulb Temperature [°C]
Risk
5.0
4.5
4.0
3.5
3.0
2.5
2.0
1.5
1.0
0.5
0.0
Inlet dP 2.5 in.H2O

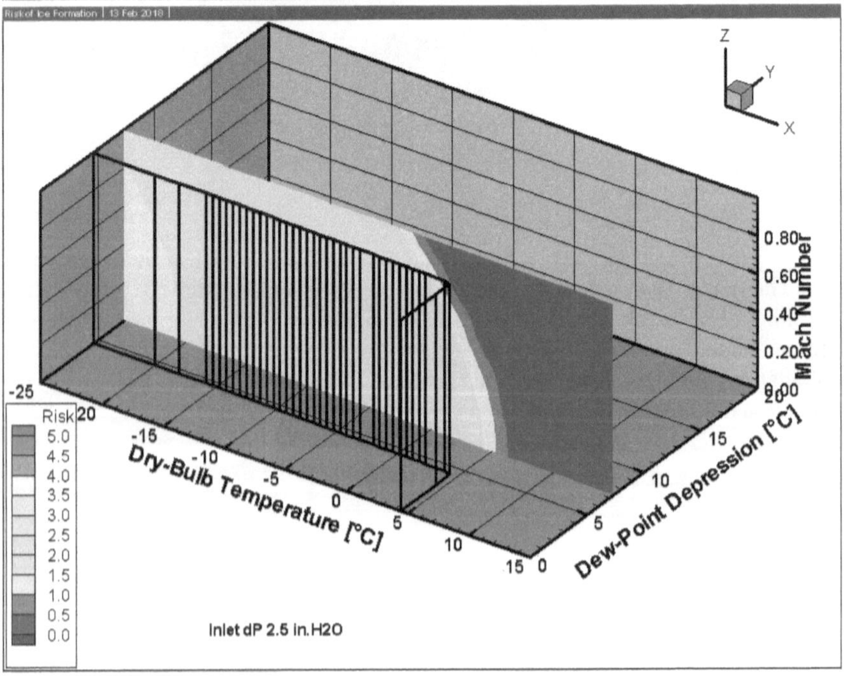
Risk of Ice Formation | 13 Feb 2018
Z
Y
X
0.80
0.60
0.40
0.20
0.00
Mach Number
-25
20
15
10
5
0
Dew-Point Depression [°C]
-15
-10
-5
0
5
10
15
Dry-Bulb Temperature [°C]
Risk
5.0
4.5
4.0
3.5
3.0
2.5
2.0
1.5
1.0
0.5
0.0
Inlet dP 2.5 in.H2O

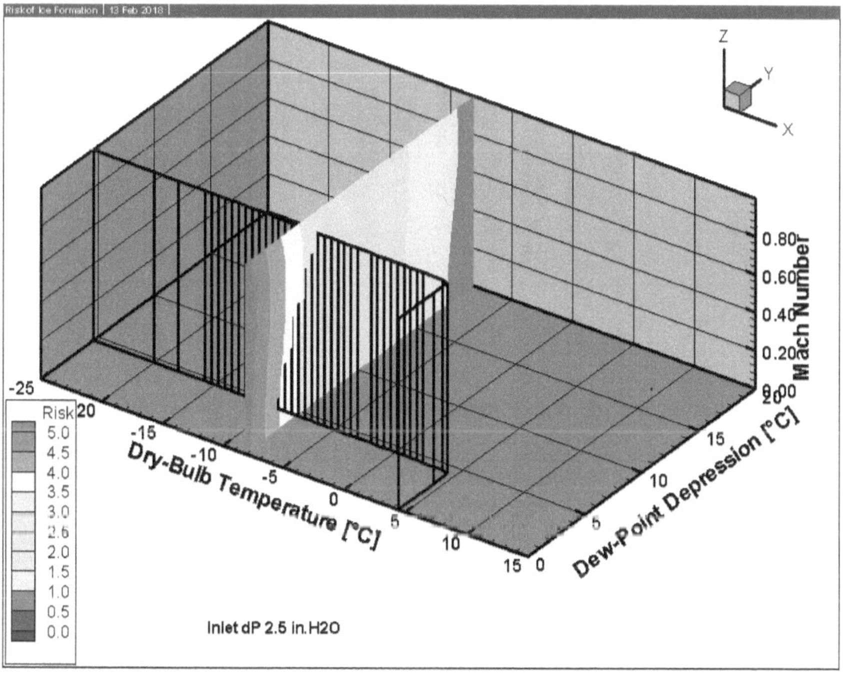

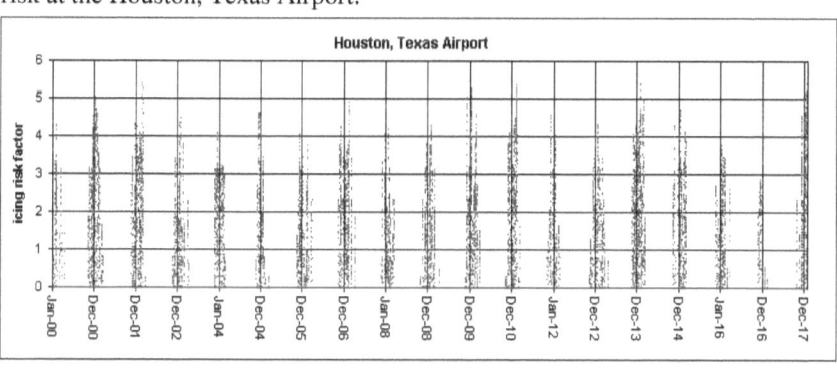

Of course, risk if icing depends on location. This next figure is shows the risk at the Houston, Texas Airport:

95

Of course, the risk at Edmonton, Alberta is much higher:

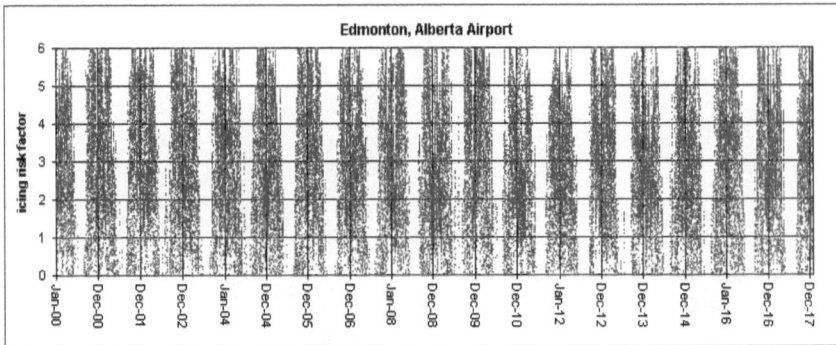

A comparison of the cumulative risk (integral of the probability) for these two locations is shown in this next figure:

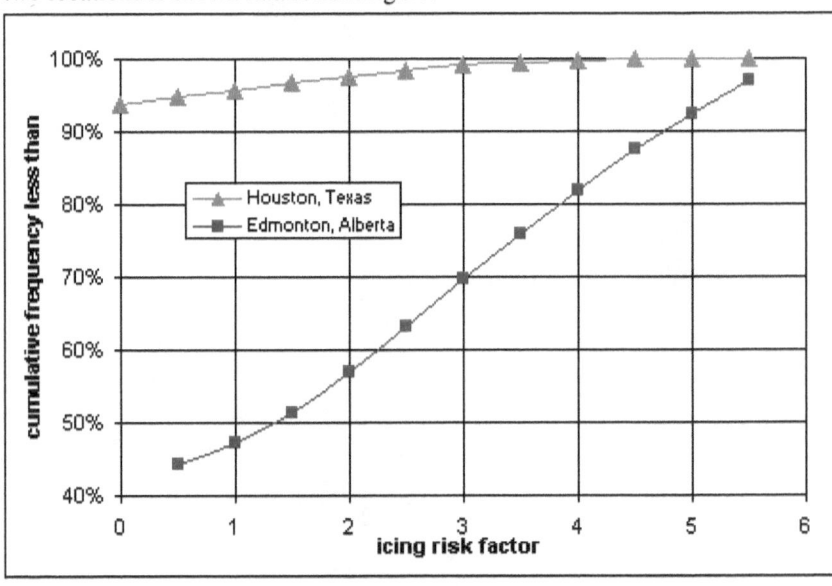

## Appendix G. Gas Turbine Heat and Mass Balance

Performing a heat balance around a gas turbine is far more complicated than it may seem. Not only are there multiple streams entering and leaving, one must also consider the chemistry. The most common mistake made in analyzing such systems is mixing enthalpies from different sources with different reference points.

The properties of moist air are defined by the American Society of Heating, Refrigeration, and Air Conditioning Engineers (ASHRAE) in their Handbook of Fundamentals, which is published periodically (Appendix A). This is the de-facto worldwide standard. The reference point for these properties is as follows: 1) The enthalpy of dry air is zero at 0°F; and 2) The enthalpy of water vapor is zero at the triple point (i.e., 32.018°F). This difference in reference points for the dry air and water vapor is not a problem as long as you are dealing with moist air near atmospheric conditions and use this same formulation throughout. The problem comes when you mix it with something else.

Properties of the fuel (whether gas or liquid) are not based on either 0°F or 32.018°F. There are many sources for fuel properties and it is important to recognize that the sensible and chemical energies may be at different reference conditions. The *CRC Handbook of Chemistry and Physics* is one common reference. Primary industrial references include the American Society of Mechanical Engineers (ASME) Power Test Code (PTC) 22 (Gas Turbines) and 4.4 (Gas Turbine Heat Recovery Steam Generators).[21]

There are three primary sources for gas (not fuel) properties: 1) NASA Glenn[22], 2) JANAF Gas Tables[23], and 3) the VDI Wärmeatlas.[24] All three are excellent. They also have different reference points, including 25°C/77°F or 0°C/32°F. To further complicate matters, there are on-line sources that may use yet different references.

---

[21] Note that these two are not exactly the same in spite of statements to that effect. Note also that both of these claim to derive properties from the NASA Glenn Report as well as (Gas Producers Association Report) GPA-2145. While this is loosely true, it is not literally accurate. If you compare the numbers listed in the ASME documents to the NASA and GPA documents, they are not the same. This is due to conversions and adjustments that have been made by ASME.

[22] McBride, B. J., Zehe, M. J., Gordon, S., "NASA Glenn Coefficients for Calculating Thermodynamic Properties of Individual Species," NASA Report No. 211556, 2002.

[23] Chase M. W., Davies, C. A., Downey, J. R., Frurip, D. J., McDonald, R. A., and Syverud, A. N., "JANAF [Joint Army, Navy, Air Force] Thermochemical Tables, Third Edition,", Journal of Physical Chemistry Reference Data, Vol. 14, Supplement 1, 1985.

[24] Mewes, D., VDI-Wärmeatlas [Heat Atlas], 8th edition. VDI [Association of German Engineers] Society for Chemical Engineering, Springer-Verlag, 1997.

A gas turbine may also have water or steam injection, which is done to control NOx emissions and increase performance. It is essential to recognize that the steam (or water) properties do not have the same reference point as the water vapor enthalpies provided by NASA Glenn, JANAF, or VDI. If you subtract the exit enthalpy for water vapor in the exhaust from the steam enthalpy of the injection, you will introduce a significant error.

To further complicate matters, the heat of combustion is also dependent on reference temperature, as noted in the following figure, excerpted from a common engineering handbook (also see Appendix H).

| Formula | State | Heat of combustion $-\Delta H_c°$, at 25°C. and constant pressure, to form | | | | | |
| --- | --- | --- | --- | --- | --- | --- | --- |
| | | H₂O (liq.) and CO₂ (gas) | | | H₂O (gas) and CO₂ (gas) | | |
| | | Kcal./mole | Cal./g. | B.t.u./lb. | Kcal./mole | Cal./g. | B.t.u./lb. |
| H₂ | gas | 68.3174 | 33,887.6 | 60,957.7 | 57.7979 | 28,669.6 | 51,571.4 |
| C | solid, graph. | 94.0518 | 7,831.1 | 14,086.8 | | | |
| CO | gas | 67.6361 | 2,414.7 | 4,343.6 | | | |
| Paraffins | | | | | | | |
| CH₄ | gas | 212.798 | 13,265.1 | 23,861 | 191.759 | 11,953.6 | 21,502 |
| C₂H₆ | gas | 372.820 | 12,399.2 | 22,304 | 341.261 | 11,349.6 | 20,416 |
| C₃H₈ | gas | 530.605 | 12,033.5 | 21,646 | 488.527 | 11,079.2 | 19,929 |
| C₃H₈ | liq.* | 526.782 | 11,946.8 | 21,490 | 484.704 | 10,992.5 | 19,774 |
| C₄H₁₀ | gas | 687.982 | 11,837.3 | 21,293 | 635.384 | 10,932.3 | 19,665 |
| C₄H₁₀ | liq.* | 682.844 | 11,748.9 | 21,134 | 630.246 | 10,843.9 | 19,506 |
| C₄H₁₀ | gas | 686.342 | 11,809.1 | 21,242 | 633.744 | 10,904.1 | 19,614 |
| C₄H₁₀ | liq.* | 681.625 | 11,727.9 | 21,096 | 629.027 | 10,822.9 | 19,468 |
| C₅H₁₂ | gas | 845.16 | 11,714.6 | 21,072 | 782.04 | 10,839.7 | 19,499 |
| C₅H₁₂ | liq. | 838.80 | 11,626.4 | 20,914 | 775.68 | 10,751.5 | 19,340 |
| C₅H₁₂ | gas | 843.24 | 11,688.0 | 21,025 | 780.12 | 10,813.1 | 19,451 |

You **must** use the same reference for all of the properties **and** the heating value; otherwise, you will introduce errors into the calculation. The following figure illustrates this problem:

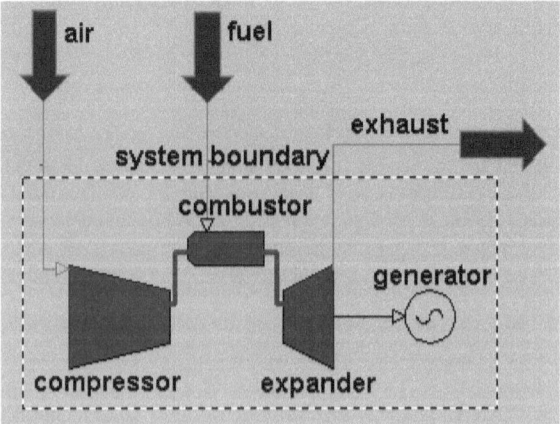

The following simplifications will be made to facilitate this example:
1) The fuel is pure methane.
2) There is no water or steam injection.

3) There is no bleed off of the compressor.

4) There is no gear loss (i.e., a dual-shaft engine).

5) The impact of fuel pressure on sensible heat is negligible.

6) The trace gases found in air (Hydrogen, , Neon, Krypton, Xenon Methane, and Sulfur Dioxide) can be lumped together with Helium, Argon, or Carbon Dioxide.

7) Complete combustion without NOx formation.

8) The heat loss can be characterized by an assumed percentage of the total heat input.

This problem is solved by first calculating the amount of oxygen required for combustion of the fuel. This has been hardwired into the Excel® spreadsheet for methane, but can be generalized using a table, as in the more complete spreadsheet, which is also included in the on-line archive. The fuel composition can also be generalized using tables (also in the more complete example).

The enthalpy is then calculated for each of the streams entering and leaving. This is done by considering the dry air required for stoichiometric combustion, the excess dry air, the moisture (water vapor) brought in with the dry air, the fuel, and each of the products of combustion as separate *streams*. These *streams* are entirely mixed, but the chemical species are considered separate and properly accounted for, so that the end result is the same. All of the enthalpies-including the moist air at the inlet-are calculated at the same reference condition: 25°C/77°F.

By considering the dry air required for combustion separate from the excess air, this adds an equation and eliminates the necessity of an iterative solution, as is the case if the combustion and excess air are considered jointly. The water vapor in the exhaust is equal to that brought in with the dry combustion air plus that brought in with the excess air plus that resulting from hydrocarbon combustion. The same is true for the carbon dioxide. The results are shown in the following figure for SI units. A second tab in the spreadsheet provides the same calculations only with English units.

Note that the enthalpy of a mixture of gases must be calculated on a mass fraction basis, not a mole fraction basis. Fuel gas composition is most often provided on a mole fraction basis, so this must also be considered when solving this problem for a mixture, such as natural gas. Also remember that the humidity ratio (mass of water vapor per mass of dry air) must be used and not relative humidity, which is a percentage of saturation and not a mass-based property.

It is also important to use what is called the *Lower Heating Value* (LHV) of the fuel and not the *Higher Heating Value* (HHV). See Appendix B for more details on LHV vs. HHV.

99

## Gas Turbine Heat Balance Calculations (Methane Fuel Only)

| Inputs | Value | Units | Calculations | Value | Units | Exhaust Mass Fractions | | Units |
|---|---|---|---|---|---|---|---|---|
| barometric pressure | 101.0 | kPa | humidity ratio | 0.015321 | kg/kg | Nitrogen | 72.9043% | - |
| ambient temperature | 35.0 | °C | shaft power | 169,591 | kWt | Carbon Dioxide | 5.4861% | - |
| ambient relative humidity | 43% | - | heat losses | 3,854 | kWt | Water (vapor) | 5.9342% | - |
| fuel flow (methane) | 9.460 | kg/s | Reactants | Value | Units | Oxygen | 14.4273% | - |
| fuel temperature | 185 | °C | Carbon Dioxide | 25.924 | kg/s | Argon | 1.2461% | - |
| exhaust temperature | 600 | °C | Water (vapor) | 21.224 | kg/s | Helium | 0.0001% | - |
| generator output | 167,700 | kWe | stoichiometric O2 | 37.699 | kg/s | total | 100.0000% | - |
| generator efficiency | 98.89% | - | dry air | 162.902 | kg/s | Enthalpies | Value | Units |
| heat loss | 0.82% | - | moist air | 165.398 | kg/s | ambient dry air | 10.05 | kJ/kg |
| fuel lower heating value | 50,000 | kJ/kg | Products | Value | Units | ambient moisture | 18.04 | kJ/kg |
| Dry Air Constituents by mole | | Units | Carbon Dioxide | 26.003 | kg/s | fuel (sensible) | 432.21 | kJ/kg |
| Nitrogen | 78.0842% | - | Water (vapor) | 21.224 | kg/s | excess air at Tex | 605.30 | kJ/kg |
| Oxygen | 20.9477% | - | Nitrogen | 123.022 | kg/s | exhaust moisture | 1144.23 | kJ/kg |
| Argon | 0.9359% | - | Argon | 2.103 | kg/s | products | 696.55 | kJ/kg |
| Carbon Dioxide | 0.0316% | - | Helium | 0.000 | kg/s | exhaust | 640.02 | kJ/kg |
| Helium | 0.0006% | - | total | 172.352 | kg/s | | | |
| total | 100.0000% | - | excess dry air | 297.001 | kg/s | | | |
| Molecular Weights | Value | Units | excess moist air | 301.552 | kg/s | | | |
| Hydrogen | 2.016 | 1/mole | total air in (moist) | 466.950 | kg/s | | | |
| Helium | 4.003 | 1/mole | total exh. (moist) | 476.400 | kg/s | | | |
| Carbon | 12.011 | 1/mole | Exhaust Constituents | | Units | | | |
| Nitrogen | 28.013 | 1/mole | Nitrogen | 347.316 | kg/s | | Legend | |
| Oxygen | 31.999 | 1/mole | Carbon Dioxide | 26.145 | kg/s | | user inputs | |
| Argon | 39.948 | 1/mole | Water (vapor) | 28.270 | kg/s | | calculated values | |
| Methane | 16.042 | 1/mole | Oxygen | 68.732 | kg/s | | | |
| Water | 18.015 | 1/mole | Argon | 5.937 | kg/s | | | |
| Carbon Dioxide | 44.010 | 1/mole | Helium | 0.000 | kg/s | | | |
| Air (dry) | 28.965 | 1/mole | total | 476.400 | kg/s | | | |

Most of the equations in this spreadsheet are fairly simple, for example, mass flows times ratios of molecular weights and heating value times fuel flow. Even the dry air required for stoichiometric combustion is straightforward. The complicated equation arises when solving the overall energy balance for the flow of excess dry air, which is provided below:

$$m_{excess\_dry} = \frac{\left( \begin{array}{c} m_{fuel} \left( LHV + h_{fuel,\ Tfuel} - h_{products,\ Texh} \right) - Q_{loss} - W_{shaft} \\ - \dfrac{\left( h_{air,\ Tbleed} - h_{air,\ Tin} + \omega \left( h_{H2O,\ Tbleed} - h_{H2O,\ Tin} \right) \right) m_{bleed\_total}}{1 + \omega} \\ - m_{comb\_dry} \left( h_{products,\ Texh} - h_{air,\ Tin} + \omega \left( h_{H2O,\ Texh} - h_{H2O,\ Tin} \right) \right) \\ - m_{inject} \left( h_{H2O,\ Texh} - h_{H2O,\ Tinj} \right) \end{array} \right)}{\left( h_{air,\ Texh} - h_{air,\ Tin} + \omega \left( h_{H2O,\ Texh} - h_{H2O,\ Tin} \right) \right)}$$

## Appendix H. Apparent Temperature Dependence of Heating Values

In order to illustrate this concept, a simple example will be presented, namely: complete, stoichiometric combustion of hydrogen with oxygen to form water. Consider the control volume shown below:

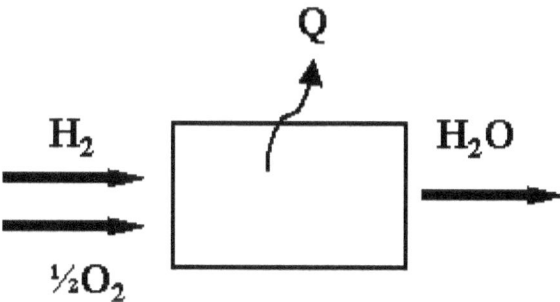

The conservation of energy for this control volume is given by the following equation:

$$\sum_{in} m_i (h_i + H_i) = Q + \sum_{exit} m_i (h_i + H_i) \qquad (H.1)$$

where $m_i$ is the mass flow rate, $h_i$ is the enthalpy, and $H_i$ is the heat of formation of each substance or stream and $Q$ is the heat transfer. The heat of formation of hydrogen and oxygen are both zero; but for water, the heat of formation is negative, which is why $Q$ is positive and this reaction releases heat. For the purposes of this illustration, the enthalpies will be approximated by the following formula:

$$h = \int_{T_{ref}}^{T} C_P dT \approx C_P (T - T_{ref}) \qquad (H.2)$$

where $C_P$ is the specific heat, $T$ is the temperature, and $T_{ref}$ is the reference temperature, typically 25°C/77°F. In this case, the mole fractions would be 2/3 for hydrogen and 1/3 for oxygen. The molecular weights are 2, 32, and 18, for hydrogen, oxygen, and water, respectively; therefore, the mass fractions would be 1/9 for hydrogen and 8/9 for oxygen. The formula for specific heat (H.2) can be substituted into the conservation of energy (H.1) to produce the following:

$$\left( \frac{C_{P,Hydrogen} + 8C_{P,Oxygen}}{9} \right)(T_{in} - T_{ref})$$

$$= \frac{Q}{m_{Water}} + \left[ C_{P,Water} (T_{exit} - T_{ref}) + H_{Water} \right] \qquad (H.3)$$

For the isothermal case, where the inlet and exit temperatures are the same, Equation H.3 can be solved for the heat of formation of water, $H_{water}$:

$$H_{Water} = \left( \frac{C_{P,Hydrogen} + 8C_{P,Oxygen} - 9C_{P,Water}}{9} \right)$$
$$\left(T - T_{ref}\right) - \frac{Q}{m_{Water}} \tag{H.4}$$

If the reaction occurs at $T=T_{ref}$, the first term on the right side goes to zero and $H_{water} = -Q/m_{water}$, which for $T_{ref}=25°C/77°F$ is 5771 BTU/lbm. Water having a negative heat of formation of 5771 BTU/lbm is equivalent to hydrogen having a positive heating value of 9 times this, or 51,939 BTU/lbm. This is straightforward enough; but consider the adiabatic case where $Q=0$ and the inlet and exit temperatures are different. Equation 3 can be solved for $T_{exit}$:

$$T_{exit} = T_{ref} + \frac{\left(C_{P,Hydrogen} + 8C_{P,Oxygen}\right)\left(T_{in} - T_{ref}\right) - 9H_{Water}}{9C_{P,Water}} \tag{H.5}$$

The specific heat is approximately 3.42, 0.22, and 0.65 BTU/lbm/°F for hydrogen, oxygen, and water, respectively. For $T_{ref}=25°C/77°F$, the exit temperature is approximately 9000°F. [The adiabatic flame temperature of hydrogen is closer to 5800°F due to dissociation.] Equation 5 can be rearranged to break out $T_{ref}$:

$$T_{exit} = \left( 1 - \frac{C_{P,Hydrogen} + 8C_{P,Oxygen}}{9C_{P,Water}} \right) T_{ref}$$
$$+ \frac{\left(C_{P,Hydrogen} + 8C_{P,Oxygen}\right)T_{in} - 9H_{Water}}{9C_{P,Water}} \tag{H.6}$$

The only time $T_{exit}$ would be independent of $T_{ref}$ is if the specific heats just happen to cause the term preceding $T_{ref}$ to be zero, which might happen for some contrived example with constant specific heats, but is unlikely to ever occur with variable specific heats. Obviously, the exit temperature is independent of the reference temperature, which means that the heat of formation of water (or the heating value of hydrogen) is associated with a specific reference temperature, most often 25°C/77°F, as is the case with the NASA Glenn data. This means that, if you want to use some other reference temperature than the basis of the tables, you must also adjust the associated heats of formation (or the heating values). Equation H.6 can be reorganized to show this adjustment for this simplified case of constant specific heats:

$$H_{ref\,2} = H_{ref\,1}$$
$$+ \left( \frac{9C_{P,Water} - C_{P,Hydrogen} - 8C_{P,Oxygen}}{9} \right) \left( T_{ref\,2} - T_{ref\,1} \right) \tag{H.7}$$

Equation H.7 also shows that the only time changing the reference temperatures wouldn't matter is if the combination of mass fractions and specific heats just happened to make the first term in parentheses equal to zero. Of course, the required adjustment to the heat of formation (or heating value) for non-ideal behavior with variable specific heats is considerably more complicated, but easily programmable.

In summary, the heat of formation (or heating value) doesn't really change with temperature; but it is associated with a specific reference temperature, due to its interdependence with the properties of the various reactants and products; however, if you want to change the reference temperature, you must also adjust the heat of formation (or heating value). This isn't a real physical change, but rather an apparent change, as the behavior of molecules is obviously independent of our reference points. So, either use the reference temperature associated with the property tables or be prepared to make adjustments accordingly. In conclusion, changing reference temperatures does not simplify any calculations.

## also by D. James Benton

*3D Articulation: Using OpenGL*, ISBN-9798596362480, Amazon, 2021 (book 3 in the 3D series).

*3D Models in Motion Using OpenGL*, ISBN-9798652987701, Amazon, 2020 (book 2 in the 3D series.

*3D Rendering in Windows: How to display three-dimensional objects in Windows with and without OpenGL*, ISBN-9781520339610, Amazon, 2016 (book 1 in the 3D series).

*A Synergy of Short Stories: The whole may be greater than the sum of the parts*, ISBN-9781520340319, Amazon, 2016.

*Azeotropes: Behavior and Application*, ISBN-9798609748997, Amazon, 2020.

*bat-Elohim: Book 3 in the Little Star Trilogy*, ISBN-9781686148682, Amazon, 2019.

*Boilers: Performance and Testing*, ISBN: 9798789062517, Amazon 2021.

*Combined 3D Rendering Series: 3D Rendering in Windows®, 3D Models in Motion, and 3D Articulation*, ISBN-9798484417032, Amazon, 2021.

*Complex Variables: Practical Applications*, ISBN-9781794250437, Amazon, 2019.

*Compression & Encryption: Algorithms & Software*, ISBN-9781081008826, Amazon, 2019.

*Computational Fluid Dynamics: an Overview of Methods*, ISBN-9781672393775, Amazon, 2019.

*Computer Simulation of Power Systems: Programming Strategies and Practical Examples*, ISBN-9781696218184, Amazon, 2019.

*Contaminant Transport: A Numerical Approach*, ISBN-9798461733216, Amazon, 2021.

*CPUnleashed! Tapping Processor Speed*, ISBN-9798421420361, Amazon, 2022.

*Curve-Fitting: The Science and Art of Approximation*, ISBN-9781520339542, Amazon, 2016.

*Death by Tie: It was the best of ties. It was the worst of ties. It's what got him killed.*, ISBN-9798398745931, Amazon, 2023.

*Differential Equations: Numerical Methods for Solving*, ISBN-9781983004162, Amazon, 2018.

*Equations of State: A Graphical Comparison*, ISBN-9798843139520, Amazon, 2022.

*Evaporative Cooling: The Science of Beating the Heat*, ISBN-9781520913346, Amazon, 2017.

*Forecasting: Extrapolation and Projection*, ISBN 9798394019494, Amazon 2023.

*Heat Engines: Thermodynamics, Cycles, & Performance Curves*, ISBN-9798486886836, Amazon, 2021.

*Heat Exchangers: Performance Prediction & Evaluation*, ISBN-9781973589327, Amazon, 2017.

*Heat Recovery Steam Generators: Thermal Design and Testing*, ISBN-9781691029365, Amazon, 2019.

*Heat Transfer: Heat Exchangers, Heat Recovery Steam Generators, & Cooling Towers*, ISBN-9798487417831, Amazon, 2021.

*Heat Transfer Examples: Practical Problems Solved*, ISBN-9798390610763, Amazon, 2023.

*The Kick-Start Murders: Visualize revenge*, ISBN-9798759083375, Amazon, 2021.

*Jamie2: Innocence is easily lost and cannot be restored*, ISBN-9781520339375, Amazon, 2016-18.

*Kyle Cooper Mysteries: Kick Start, Monte Carlo, and Waterfront Murders*, ISBN-9798829365943, Amazon, 2022.

*The Last Seraph: Sequel to Little Star*, ISBN-9781726802253, Amazon, 2018.

*Little Star: God doesn't do things the way we expect Him to. He's better than that!* ISBN-9781520338903, Amazon, 2015-17.

*Living Math: Seeing mathematics in every day life (and appreciating it more too)*, ISBN-9781520336992, Amazon, 2016.

*Lost Cause: If only history could be changed...*, ISBN-9781521173770, Amazon, 2017.

*Mass Transfer: Diffusion & Convection*, ISBN-9798702403106, Amazon, 2021.

*Mill Town Destiny: The Hand of Providence brought them together to rescue the mill, the town, and each other*, ISBN-9781520864679, Amazon, 2017.

*Monte Carlo Murders: Who Killed Who and Why*, ISBN-9798829341848, Amazon, 2022.

*Monte Carlo Simulation: The Art of Random Process Characterization*, ISBN-9781980577874, Amazon, 2018.

*Nonlinear Equations: Numerical Methods for Solving*, ISBN-9781717767318, Amazon, 2018.

*Numerical Calculus: Differentiation and Integration*, ISBN-9781980680901, Amazon, 2018.

*Numerical Methods: Nonlinear Equations, Numerical Calculus, & Differential Equations*, ISBN-9798486246845, Amazon, 2021.

*Orthogonal Functions: The Many Uses of*, ISBN-9781719876162, Amazon, 2018.

*Overwhelming Evidence: A Pilgrimage*, ISBN-9798515642211, Amazon, 2021.

*Particle Tracking: Computational Strategies and Diverse Examples*, ISBN-9781692512651, Amazon, 2019.

*Plumes: Delineation & Transport*, ISBN-9781702292771, Amazon, 2019.

*Practical Linear Algebra: Principles & Software*, ISBN-9798860910584, Amazon, 2023.

*Props, Fans, & Pumps: Design & Performance*, ISBN-9798645391195, Amazon, 2020.

*Remediation: Contaminant Transport, Particle Tracking, & Plumes*, ISBN-9798485651190, Amazon, 2021.

*ROFL: Rolling on the Floor Laughing*, ISBN-9781973300007, Amazon, 2017.

*Seminole Rain: You don't choose destiny. It chooses you*, ISBN-9798668502196, Amazon, 2020.

*Septillionth: 1 in $10^{24}$*, ISBN-9798410762472, Amazon, 2022.

*Software Development: Targeted Applications*, ISBN-9798850653989, Amazon, 2023.

*Software Recipes: Proven Tools*, ISBN-9798815229556, Amazon, 2022.

*Steam 2020: to 150 GPa and 6000 K*, ISBN-9798634643830, Amazon, 2020.

*Thermochemical Reactions: Numerical Solutions*, ISBN-9781073417872, Amazon, 2019.

*Thermodynamic and Transport Properties of Fluids*, ISBN-9781092120845, Amazon, 2019.

*Thermodynamic Cycles: Effective Modeling Strategies for Software Development*, ISBN-9781070934372, Amazon, 2019.

*Thermodynamics - Theory & Practice: The science of energy and power*, ISBN-9781520339795, Amazon, 2016.

*Version-Independent Programming: Code Development Guidelines for the Windows® Operating System*, ISBN-9781520339146, Amazon, 2016.

*The Waterfront Murders: As you sow, so shall you reap*, ISBN-9798611314500, Amazon, 2020.

*Weather Data: Where To Get It and How To Process It*, ISBN-9798868037894, Amazon, 2023.